T0215015

Periodic Solutions of First-Order Functional Differential Equations in Population Dynamics

Seshadev Padhi · John R. Graef
P. D. N. Srinivasu

Periodic Solutions of First-Order Functional Differential Equations in Population Dynamics

Springer

Seshadev Padhi
Department of Applied Mathematics
Birla Institute of Technology, Mesra
Ranchi, Jharkhand
India

P. D. N. Srinivasu
Department of Mathematics
Andhra University
Visakhapatnam, Andhra Pradesh
India

John R. Graef
Department of Mathematics
University of Tennessee at Chattanooga
Chattanooga, TN
USA

ISBN 978-81-322-3542-2   ISBN 978-81-322-1895-1   (eBook)
DOI 10.1007/978-81-322-1895-1
Springer New Delhi Heidelberg New York Dordrecht London

Dedicated to my father and mother:
Sri Bansidhar Padhi and Smt. Damayanti
Padhi

Seshadev Padhi

Dedicated to my wife Frances

John R. Graef

Dedicated to Bhagawan Sri Sathya Sai Baba,
the founder Chancellor of my parent
University, Sri Sathya Sai Institute
of Higher Learning

P. D. N. Srinivasu

Preface

Ordinary differential equations play a crucial role in providing answers to real-world problems and they continue to be an indispensable tool in scientific investigations. Although these equations are often first approximations to real-world systems, they can be refined into more accurate models by including information about past states in time into the equations, e.g., formulating them as delay differential equations, or more generally, functional differential equations. The rapid progress observed in this area is mainly due to their ability to capture the dynamics observed in real-word phenomena.

The literature on ordinary and delay differential equations is large and rapidly growing; results appear in a variety of journals, not just those in pure mathematics. This is due in part to their wide applicability. The domain of investigation and the type of contribution that is needed is often dictated by the application to be modeled. For example, it is most appropriate to study the existence and stability behavior of positive periodic solutions of an ecological system that is strongly influenced by periodic environmental variations.

The capability of functional differential equations to mimic the dynamics in a periodically fluctuating environment has been the driving force for researchers to make valuable contributions to such problems. There has also been great interest in finding conditions under which periodic functional differential equations admit multiple positive periodic solutions. Such problems have wide applications in biology.

In this context, models that attract the attention of researchers include the Hematopoiesis (red blood cell production) model, the Lasota-Wazewska model, Nicholson's Blowflies model, and models with Allee effects. These models have their roots in population dynamics and are extensively used to describe real-world problems. To the best of our knowledge, there is no book that systematically covers the dynamics and global aspects of these models in a periodic environment. Contributions to this area are quite recent. The necessity to bring together the theory and applications is the motivation for this monograph.

The basic tool we use here to prove the existence of multiple periodic solutions is the well-known Leggett-Williams multiple fixed point theorem. This theorem has been applied to problems on the existence of multiple solutions to boundary value problems. The approach is to transform the problem into an equivalent integral equation from which an integral operator is easily formed. The fixed points of the operator then correspond to solutions of the original problem. Of course, an appropriate mathematical setting for the operator must be constructed.

A brief description of the organization of this monograph is as follows; there are a total of five chapters. In Chap. 1, we introduced the Leggett-Williams multiple fixed point theorems that guarantee the existence of two and three fixed points of operators. These fixed point theorems are used throughout the book in showing the existence of two or three positive periodic solutions of various equations and models.

Chapter 2 is concerned with the existence of at least three positive periodic solutions of first-order nonlinear functional differential equations. The results are applied to several population models, such as those listed above. Chapter 3 presents sufficient conditions for the existence of at least three positive periodic solutions for a system of nonlinear functional differential equations. An application to the Hematopoiesis model highlights the significance of the delay in the model.

Many interesting and easily verifiable conditions on the coefficient functions are given in Chaps. 2 and 3 for the existence of at least three positive periodic solutions of first-order nonlinear functional differential equations where the nonlinear terms are normally unimodal. This excludes some interesting population models whose nonlinear term is either non-decreasing or non-increasing. Thus, an attempt has been made in Chap. 4 to find existence of periodic solutions for nonlinear equations with non-decreasing nonlinear term and established existence of at least two positive periodic solutions.

Chapter 5 concentrates on the existence of unique positive periodic solutions and their global asymptotic stability. Results on the existence of a unique globally asymptotically stable periodic solution for the fishing model, the Hematopoiesis model, Nicholson's Blowflies model, and the Lasota-Wazewska model are presented.

This book contains a large number of references and we hope this will be useful to readers in their future research work.

We owe thanks to many people who have rendered their help during the preparation of this monograph. In particular we wish to thank Chuanxi Qian (Mississippi State University, USA), Julio G. Dix (Texas State University at San Marcos, USA), Jaffar Ali Shahul Hameed (Florida Gulf Coast University, USA), N. C. Mahanti (BIT, Mesra, India), and E. Thandapani (RIASM, Chennai, India) for their inspiration and encouragement. We are thankful to Suchitra, Vasantha Kumari, Sai Aneesha, Sai Anurag and Aryadev for their constant inspiration during the preparation of this monograph. We also wish to express our appreciation to Smita Pati for her reading of the manuscript and her helpful suggestions. We are, especially, thankful to Aryadev for his careful proof reading of this book. We should also mention here that it was a great pleasure to work with Mr. Shamim Ahmad and Ms. Nupoor Singh, the editorial team of Springer, and the production team, especially, to Ayswaraya N. who took utmost care during the preparation of the book. Finally, we gratefully acknowledge B. S. R. V. Prasad for his help in the preparation of this monograph.

Seshadev Padhi
John R. Graef
P. D. N. Srinivasu

Contents

About the Authors

Seshadev Padhi is associate professor of Mathematics at Birla Institute of Technology, Mesra, Ranchi, Jharkhand, India. He received his Ph.D. on the topic "Oscillation Theory of Third Order Differential Equations." He was awarded the BOYSCAST (Better Opportunities for Young Scientists in Chosen Areas of Science and Technology) fellow by the Department of Science and Technology (DST), Government of India in 2004 to visit Mississippi State University, USA. Subsequently, Dr. Padhi did his post-doctoral work at Mississippi State University, USA. In addition, Dr. Padhi visited several Institutes of international repute: Florida Institute of Technology, Melbourna, Florida USA to work in Collaboration with Prof. T. Gnanabhaskar in 2006; Texas State University at San Marcos, Texas, USA to work in collaboration with Prof. Julio G. Dix, in 2009; University of Tennessee at Chattanooga, Chattanooga, Tennessee, USA in 2011–2013 to work in collaboration with Prof. John R. Graef; University of Szeged, Szeged, Hungary in 2007 and 2011 to work in collaboration with Prof. Tibor Krisztin. Besides, he also visited ETH, Zurich, Switzerland under Borel Set Theory Programme in 2005, and also several countries to deliver lectures in different international conferences and workshops. Dr. Padhi got UNESCO travel and lodging grant to visit ICTP, Trieste, Italy in the year 2003. Dr. Padhi has published more than 60 research papers in international journals of repute. He has been working as referee for more than 30 international journals and a reviewer of Mathematical Review since 2006.

John R. Graef is professor of Mathematics at the University of Tennessee at Chattanooga. His research interests include ordinary and functional differential equations, difference equations, impulsive systems, differential inclusions, dynamic equations on time scales, fractional differential equations, and their applications. His special interests are in nonlinear oscillations, stability and other asymptotic properties of solutions, boundary value problems, and applications to biological systems. He has published more than 350 papers and authored or edited five monographs.

P. D. N. Srinivasu is professor of Mathematics at Andhra University, Visakhapatnam, India. He obtained his Ph.D. in 1992 from Sri Sathya Sai Institute of Higher Learning, India, on the topic "Existential and numerical study of implicit differential equations." His research interests include mathematical modeling, population dynamics, mathematical bio-economics, and optimal control. He visited several academic institutes of international repute to deliver invited lectures and participate in conferences and workshops. Prof. Srinivasu visited The International Centre for Theoretical Physics (ICTP), Trieste, Italy, many times as a visiting scientist, in addition to The Beijer International Institute of Ecological Economics, Royal Swedish Academy of Sciences, Stockholm, Sweden; Eidgenossische Technische Hochschule (ETH), Zurich; and the University of Zurich, Switzerland. Besides working as a reviewer for many internationally journals, Prof. Srinivasu has many research papers to his credit in several journals of repute.

Chapter 1
Introduction

Qualitative theory of differential equations deals with the behavior of solutions without finding them explicitly. Existence, uniqueness, oscillation, nonoscillation, asymptotic behavior, periodicity, stability, asymptotic stability, and global attractivity of solutions are studied.

One characteristic phenomenon of population dynamics is the often observed oscillation behavior of the population densities. To better understand such a phenomenon, one mechanism is to introduce time delays in the models, which results in models described by functional differential equations. A functional differential equation is a more general type of differential equation in which the unknown function occurs at various different times. For example,

$$x'(t) = f(t, x(t), x(t - \tau(t))), \quad 0 < \tau(t) < t$$

is a first-order delay differential equation,

$$x'(t) = f(t, x(t), x(t + \tau(t))), \quad 0 < \tau(t)$$

is a first-order advanced differential equation, and

$$x'(t) = f(t, x(t), x(t - \tau(t)), x'(t - \tau(t))), \quad 0 < \tau(t) < t$$

is a first-order neutral delay differential equation. The same definition can be applied to higher order functional differential equations and system of differential equations. The simplest and perhaps most natural type of functional differential equation is a delay differential equation. The salt brine example in Driver [16] shows how a first-order delay differential equation occurs in our day to day life. Similarly, the predator-prey model in Driver [16] shows the existence of a system of first- order delay differential equations in ecology. The monograph of Driver [16], E'lsgolts and Norkin [17], and Hale et al. [26, 27] contain detailed elementary studies of the existence, nonexistence, stability, and asymptotic stability of solutions of first-order functional differential equations and of system of functional differential equations. One of the important and elementary methods of finding solution of a delay differential equation is the method of steps [16, 17]. Many examples arise in physical, biological, and ecological systems in which the present rate of change of some unknown function(s) depends upon some past values of the same function(s) [16, 41]. In [41], Murray shows how delay differential equation models are capable of generating limit cycle periodic solutions.

In the following, we state a global existence and uniqueness theorem for delay differential equations; for additional details, we refer the reader to [16, 17].

Consider the delay differential equation

$$x'(t) + f(t, x(t), x(t - \tau_1(t)), \ldots, x(t - \tau_n(t))) = 0, \qquad (1.1)$$

where for some $t_0 \in R$,

$$f \in C([t_0, \infty), R) \text{ and } \tau_i \in C([t_0, \infty), R^+), \quad \text{for } i = 1, 2, \ldots, n, \qquad (1.2)$$

and

$$\lim_{t \to \infty} (t - \tau_i(t)) = \infty, \quad i = 1, 2, \ldots, n.$$

For every initial point $t_1 \geq t_0$, we define $t_{-1} = t_{-1}(t_0)$ to be

$$t_{-1} = \min_{1 \leq i \leq n} \inf_{t \geq t_0} (t - \tau_i(t)). \qquad (1.3)$$

The interval $[t_{-1}, t_0]$ is called the initial interval associated with the initial point t_0, and the delay differential equation (1.1). With Eq. (1.1) and given initial point $t_1 \geq t_0$, we associate an initial condition

$$x(t) = \phi(t), \quad t_{-1} \leq t \leq t_1, \qquad (1.4)$$

where

$$\phi : [t_{-1}, t_1] \to R$$

is the initial function. A function x is said to be a *solution* of the problem (1.1) and (1.4), if x is a solution of (1.1) satisfying (1.4). In addition to conditions (1.2) and (1.3), we assume that there exists a function $p \in C([t_0, \infty), R^+)$ such that for all $t \geq t_0$ and for all $x_i, y_i \in R, i = 1, 2, \ldots, n$, the function f satisfies the global Lipschitz condition

$$\|f(t, x_0, x_1, \ldots, x_n) - f(t, y_0, y_1, \ldots, y_n)\| \leq p(t) \sum_{i=0}^{n} \|x_i - y_i\|.$$

Let $t_1 \geq t_0$ and $\phi \in C[[t_{-1}, t_1], R]$ be given. Then the initial value problem (1.1) and (1.4) has exactly one solution in the interval $[t_1, \infty)$.

Now, we consider the first-order nonlinear functional differential equations of the form

$$x'(t) = -a(t)x(t) + \lambda f(t, x(h(t))) \tag{1.5}$$

and

$$x'(t) = a(t)x(t) - \lambda f(t, x(h(t))), \tag{1.6}$$

where λ is a positive parameter, $a, h \in C(R, R_+)$, a is T-periodic, T is a positive constant, $R = (-\infty, \infty)$, $R_+ = [0, \infty)$, $a(t) \neq 0$, and $f \in C(R \times R_+, R_+)$ is periodic with period T in its first variable.

If $h(t) = t - \tau(t)$, $\tau \in C(R, R_+)$, and $0 \leq \tau(t) \leq t$, then Eqs. (1.5) and (1.6) become

$$x'(t) = -a(t)x(t) + \lambda f(t, x(t - \tau(t))) \tag{1.7}$$

and

$$x'(t) = a(t)x(t) - \lambda f(t, x(t - \tau(t))), \tag{1.8}$$

respectively.

Functional differential equations of the form (1.7) and (1.8) include many mathematical, ecological, and biological models (directly or after some transformations), such as:

(a) Lasota-Wazewska model

$$x'(t) = -a(t)x(t) + p(t)e^{-\gamma(t)x(t-\tau(t))}; \tag{1.9}$$

(b) Hematopoiesis model or model for blood cell production

$$x'(t) = -a(t)x(t) + b(t)\frac{x(t - \tau(t))}{1 + x^n(t - \tau(t))}; \tag{1.10}$$

(c) Nicholson's Blowflies model

$$x'(t) = -a(t)x(t) + b(t)x(t - \tau(t))e^{-\gamma(t)x(t-\tau(t))}; \tag{1.11}$$

(d) Logistic equation of multiplicative type with several delays

$$x'(t) = x(t)\left[a(t) - \prod_{i=1}^{n} b_i(t)x(t - \tau_i(t))\right]; \tag{1.12}$$

(e) Generalized Richards single species growth model

$$x'(t) = x(t) \left[a(t) - \left(\frac{x(t - \tau(t))}{E(t)} \right)^{\theta} \right];$$ (1.13)

(f) Generalized Michaelis-Menton type single species growth model

$$x'(t) = x(t) \left[a(t) - \sum_{i=1}^{n} \frac{b_i(t)x(t - \tau_i(t))}{1 + c_i(t)x(t - \tau_i(t))} \right];$$ (1.14)

where a, b, p, E, b_i, τ, and $\tau_i \in C(R, R_+)$, $1 \le i \le n$, are T-periodic functions.

The variation of environment plays an important role in many biological and ecological systems. In particular, the effects of a periodically varying environment are important for evolution theory as the selective forces on ecosystems in a fluctuating environment differ from those in a stable environment. Thus, it is reasonable to study the existence and global attractivity of periodic solutions of mathematical models occurring in biology and ecology. From the biological and ecological points of view, only positive solutions are important.

Functional differential equations of the form (1.5) and (1.7) can be interpreted as the standard Malthus population model $x'(t) = -a(t)x$ subject to a perturbation with periodic delay. One important question is whether these periodic functional differential equations (both the general functional differential equations and the mathematical models) can support any positive periodic solution. This is one of the rapidly growing research areas at this time.

In the past few years, there has been an increasing interest in studying dynamical characteristics, such as stability, persistence, and global attractivity of periodic solutions of functional differential equations of the forms (1.5)–(1.8) and system of functional differential equations of the form

$$x'(t) = A(t, x(t))x(t) + \lambda f(t, x_t),$$ (1.15)

where A is a diagonal $n \times n$ matrix, whose entries depend on t and on the unknown function $x = (x_1, x_2, \ldots, x_n)^T$, and $f \in C(R \times R_+^n, R_+)$ is T-periodic in t.

Many authors have studied the existence of at least one or two positive periodic solutions of (1.5)–(1.8); for example, see [12, 30, 31, 59] and the references therein. The method used in the above references are mainly the upper lower solution method [15], where the existence of at least one solution is obtained if there exists a pair of upper and lower solutions of the considered differential equation.

Krasnosel'skii's fixed point theorem [33], fixed point theorems of cone expansion and cone compression, and fixed point index theory [15, 33] have been used in the papers [12, 31, 51, 52, 58, 59] to show the existence of at least one and at least two positive periodic solutions of functional differential equations of the forms (1.5)–(1.8).

Although the above-mentioned fixed point theorems have been widely used in the literature for the existence of one or two positive periodic solutions of functional differential equations, the application of the Leggett-Williams multiple fixed point theorem [37] is relatively new to these studies. This motivates us to use it to show the existence of at least two and at least three positive periodic solutions of the functional differential equations under consideration. The sufficient conditions obtained in this work are different from existing results in the literature that have been obtained by other methods.

It is well known that the Leggett-Williams multiple fixed point Theorems 1.2.1 and 1.2.2 have been used by many authors to show the existence of multiple solutions of boundary value problems. Once a problem is transformed into an equivalent integral operator, then it is easy to study the existence of fixed points of the operator by using different fixed point theorems; the fixed points correspond to periodic solution of the problem. The Leggett-Williams theorems are used throughout this book in showing the existence of two or three positive periodic solutions.

In [53], Wang introduced the notations i_0 and i_∞ and used fixed point index theory [15] for the existence and nonexistence of i_0 and i_∞ number of positive periodic solution of the first-order functional differential equation

$$x'(t) = a(t)g(x(t))x(t) - \lambda b(t) f(x(t - \tau(t))), \qquad (1.16)$$

where λ is a positive parameter, $a, b \in C(R, [0, \infty))$ are T-periodic functions, $\int_0^T a(t) \, dt > 0$, $\int_0^T b(t) \, dt > 0$ and $\tau \in C(R, R)$ is a T-periodic function. The functions f and g satisfy the property: $f, g \in C(R_+, R_+)$, $0 < l \le g(x) \le L < \infty$ for $x \ge 0$, l, L are positive constants, and $f(x) > 0$ for $x > 0$.

Bai and Xu [2] used the Leggett-Williams multiple fixed point theorem [37] to obtain sufficient conditions for the existence of at least three nonnegative periodic solutions of (1.16). There are several research papers in the literature on the existence of positive periodic solutions of (1.16), where the boundedness condition $0 < l \le g(x) \le L < \infty$ has been used.

Now, we assume that the coefficient functions and the delay term $\tau(t)$ in the models (1.9)–(1.14) are positive and T-periodic. Different sufficient conditions have been obtained in the literature for the existence of at least one and two positive periodic solutions of (1.9)–(1.14). In studying the existence of positive periodic solutions of (1.5)–(1.8), Graef et al. [23] proved that (1.9) has always at least one positive T-periodic solution. Brouwers fixed point theorem [15] was used to prove their result. Similar results using Krasnosel'skii's fixed point theorem can be found in [51, 56, 59] and the references cited therein.

Zhang et al. [59] used fixed point theory on cones to prove that if $b(t) > a(t)$, then (1.10) has at least one positive T-periodic solution. On the other hand, Krasnosel'skii's fixed point theorem [33] has been used in [51, 57] to prove a similar result for the existence of at least one positive T-periodic solution of (1.5). Wu et al. [55] proved that the condition $b(t) > a(t)$ is a necessary and sufficient condition for the existence of at least one positive T-periodic solution of (1.10).

As pointed out earlier for the functional differential equations (1.5)–(1.8) and (1.15), it seems that the existence of at least two or three positive periodic solutions of (1.9)–(1.14) are relatively scarce in the literature.

Global attractivity of periodic solutions of mathematical models in biology and ecology plays an important role in nature. The global attractivity of solutions to a periodic solution implies that every solution of the equation converges to the periodic solution. This shows that the system is properly controlled. One may refer to [6, 7, 21–25, 31, 32, 34, 35, 38, 39, 41, 44, 46–48, 50, 54] for results on the global attractivity of solutions of first-order differential equations and some mathematical models.

1.1 An Introduction to Allee Effects

Allee effects refer to a population that has a maximal per capita growth rate at intermediate densities. This occurs when the per capita growth rate increases as density increases and decreases after the density passes a certain critical value. The logistic equation was based on the assumption that the density has a negative effect on the per-capita growth rate. However, some species often cooperate among themselves in their search for food and to escape from their predators. For example, some species form hunting groups (packs, prides, etc.) to enable them to capture large prey. Fish and birds often form schools and flocks as a defense against their predators. Some parasitic insects aggregate so that they can overcome the defense mechanisms of a host. A number of social species such as ants, termites, bees, etc., have developed complex cooperative behavior involving the division of labor, altruism, etc. Such cooperative processes have a positive feedback influence since individuals have been provided a greater chance to survive and reproduce as density increase. Aggregation and associated cooperative and social characteristics among members of a species were extensively studied in animal populations by Allee [1]. The phenomenon in which reproduction rates of individuals decrease when the density drops below a certain critical level is now known as the Allee effect.

When the density of a population becomes too large, the positive feedback effect of aggregation and cooperation may then be dominated by density-dependent stabilizing negative feedback effects due to intraspecific competition due to excessive crowding and the ensuing shortage of resources. For example, consider the delay Lotka-Volterra type single species population growth model

$$x'(t) = x(t)[a + bx(t - \tau) - cx^2(t - \tau)], \tag{1.17}$$

where $a > 0$, $c > 0$, $\tau > 0$, and b are real constants. If $\tau = 0$, the per capita growth rate is $g(x) = a + bx - cx^2$. If $b > 0$, then $g'(0) = b > 0$, and $g(x)$ achieves its maximum at $x = b/2c$, thus exhibiting an Allee effect. If $b < 0$, then $g(x)$ is a decreasing function of x for $x \geq 0$, and thus there is no Allee effect. Note that (1.17) can be interpreted as a single species model with a quadratic per capita growth rate, which represents a nonlinear approximation of more general types of nonlinear growth rates with single humps.

Allee effects have been extensively studied in recent years largely because of their potential role in the extinction of already endangered, rare, or dramatically declining species [3, 18, 49]. The Allee effect refers to a decrease in population growth rate at low population densities [4, 10, 14, 18, 19, 40, 42, 45]. There are several mechanisms that create Allee effects in populations, and a classification of these effects is presented in [5]. Some real-world examples exhibiting Allee effects can be found in [9, 14, 20, 28, 29, 36, 43]. A critical review of single species models subjected to Allee effects is presented in [8]. An equation representing the growth of a species with Allee effects and the associated dynamics are discussed in [13, 32]. Many of the discussions in the literature deal with differential equations with constant coefficients. Although seasonality is known to have considerable impact on the species dynamics, to our knowledge there does not exist much literature that discusses the dynamics of a renewable resource subjected to Allee effects in a seasonally varying environment. In this book, we introduce seasonality into the resource dynamic equation by assuming the involved coefficients to be periodic as in [11]. Our interest is to find an estimate on the number of positive periodic solutions admitted by the considered model.

Let us consider the following equation representing dynamics of a renewable resource y, that is subject to Allee effects [13, 32]:

$$\frac{dy}{dt} = ay(y - b)(c - y), \quad a > 0, \ 0 < b < c, \tag{1.18}$$

where the constants a, c, and b represent, respectively, the intrinsic growth rate, the carrying capacity of the resource, and the threshold value below which the growth rate of the resource is negative. It is well known that Eq. (1.18) admits two positive solutions given by $y(t) = b$ and $y(t) = c$ and one trivial solution as its equilibrium solutions. Equation (1.18) can be nondimensionalized to reduce the number of parameters to obtain

$$\frac{dy}{dt} = y(y - \beta)(1 - y), \quad 0 < \beta < 1. \tag{1.19}$$

Since we are interested in the dynamics of a renewable resource in a seasonally varying environment, we assume the coefficients a, b, c to be nonnegative T-periodic functions of the same period, and study the existence of T-periodic solutions. Thus, we consider

$$\frac{dy}{dt} = a(t)y(y - b(t))(c(t) - y) \tag{1.20}$$

where the nonnegative functions $c(t)$ and $b(t)$ stand for the seasonal-dependent carrying capacity and the threshold function of the species, respectively, and satisfy

$$0 < b(t) < c(t). \tag{1.21}$$

Here, $a(t)$ represents time-dependent intrinsic growth rate of the resource. Clearly, the trivial solution $(y(t) \equiv 0)$ is a periodic solution of (1.20). Since this study deals with resource dynamics, we are interested in the existence of positive periodic solutions of the equation.

The transformation

$$y(t) = c(t)x(t) \tag{1.22}$$

transforms Eq. (1.20) into

$$\frac{dx}{dt} = -\left(a(t)c^2(t)k(t) + \frac{c'(t)}{c(t)}\right)x + a(t)c^2(t)\left((1+k(t)) - x\right)x^2, \tag{1.23}$$

where

$$k(t) = \frac{b(t)}{c(t)} < 1. \tag{1.24}$$

Note that (1.23) is a particular case of a general scalar differential equation of the form

$$\frac{dx}{dt} = -A(t)x(t) + f(t, x(t)) \tag{1.25}$$

where $A \in C(R, R)$ and $f \in C(R \times R, R)$ satisfy $A(t + T) = A(t)$ and $f(t + T, x) = f(t, x)$.

1.2 Preliminaries

In the following, we list some results for ready reference.

Let X be a real Banach space. A closed convex set $K \subset X$ is called a (positive) *cone* if the following conditions are satisfied:

(i) if $x \in K$, then $\lambda x \in K$ for $\lambda \geq 0$;
(ii) if $x \in K$ and $-x \in K$, then $x = 0$.

By a *completely continuous* map, we mean a continuous function that takes bounded sets into relatively compact sets. A notion central to the Leggett-William theorem is that of a concave positive functional on a cone K, which we define as follows.

Definition 1.2.1 A continuous map $\psi : K \rightarrow [0, \infty)$ is said to be a continuous concave positive functional on K if

$$\psi(\mu x + (1 - \mu)y) \geq \mu\psi(x) + (1 - \mu)\psi(y), \quad x, y \in K, \ \mu \in [0, 1].$$

For example, if x_0 is an interior element of K, then the map $\psi : K \rightarrow [0, \infty)$ defined by $\psi(x) = \max\{\gamma : \gamma x_0 \leq x\}$ is a concave positive functional on K.

It is possible to define different types of positive concave functionals. In Chaps. 2–4, we have considered a particular positive concave functional that is the best fit to prove the results. For some other concave functionals, we refer the reader to the paper of Leggett and Williams [37].

For $a > 0$, define $K_a = \{x \in K : ||x|| < a\}$. Then $\overline{K}_a = \{x \in K : ||x|| \leq a\}$. Let $b > 0$ and $c > 0$ be constants and K and X be defined as above. Define

$$K(\psi; b, c) = \{x \in K : \psi(x) \geq b, ||x|| \leq c\}.$$

With this, we now state the following two fixed point theorems for our use in the sequel.

Theorem 1.2.1 ([37, Theorem 3.5]) *Let $c_3 > 0$ be a constant. Assume that $A : \overline{K}_{c_3} \to K$ is completely continuous, there exists a concave nonnegative functional ψ with $\psi(x) \leq ||x||$, $x \in K$, and there exist numbers c_1 and c_2 with $0 < c_1 < c_2 < c_3$ satisfying the following conditions:*

(i) $\{x \in K(\psi, c_2, c_3) : \psi(x) > c_2\} \neq \phi$ *and* $\psi(Ax) > c_2$ *if* $x \in K(\psi, c_2, c_3)$;
(ii) $||Ax|| < c_1$ *if* $x \in \overline{K}_{c_1}$;
and
(iii) $\psi(Ax) > \frac{c_2}{c_3}||Ax||$ *for each* $x \in \overline{K}_{c_3}$ *with* $||Ax|| > c_3$.

Then A has at least two fixed points x_1, x_2 in $\overline{K}_{c_3}$. Furthermore, $||x_1|| \leq c_1 < ||x_2|| < c_3$.

Theorem 1.2.2 ([37, Theorem 3.3]) *Let $X = (X, ||.||)$ be a Banach space and $K \subset X$ be a cone, and $c_4 > 0$ be a constant. Suppose there exists a concave nonnegative continuous function ψ on K with $\psi(x) \leq ||x||$ for $x \in \overline{K}_{c_4}$ and let $A : \overline{K}_{c_4} \to \overline{K}_{c_4}$ be a continuous compact map. Assume that there are numbers c_1, c_2 and c_3 with $0 < c_1 < c_2 < c_3 \leq c_4$ such that*

(i) $\{x \in K(\psi, c_2, c_3) : \psi(x) > c_2\} \neq \phi$ *and* $\psi(Ax) > c_2$ *for all* $x \in K(\psi, c_2, c_3)$;
(ii) $||Ax|| < c_1$ *for all* $x \in \overline{K}_{c_1}$;
(iii) $\psi(Ax) > c_2$ *for all* $x \in K(\psi, c_2, c_4)$ *with* $||Ax|| > c_3$.

Then A has at least three fixed points x_1, x_2 and x_3 in $\overline{K}_{c_4}$. Furthermore, we have $x_1 \in \overline{K}_{c_1}$, $x_2 \in \{x \in K(\psi, c_2, c_4) : \psi(x) > c_2\}$, and $x_3 \in \overline{K}_{c_4} \backslash \{K(\psi, c_2, c_4) \cup \overline{K}_{c_1}\}$.

1.3 Outline of the Book

The results in this book have been divided into five chapters. This chapter contains concepts that will be used throughout the remainder of this book. All elementary results needed in the remaining chapters of the monograph are incorporated as needed. The Leggett-Williams fixed point theorems stated above are the basic tools to be used.

Chapter 2 deals with the existence of at least three positive T-periodic solutions of functional differential equations of the form (1.5)–(1.8) and

$$x'(t) = \pm a(t)x(t) \mp \lambda b(t) f(t, x(h(t))), \qquad (1.26)$$

where $b \in C(R, R_+)$ is periodic with period T. The Leggett-Williams multiple fixed point theorem ([37, Theorem 3.3]), that is, Theorem 1.2.2, is used to prove the results. Some explicit intervals on the parameter λ are obtained for the existence of at least three positive T-periodic solutions of the equations. As an outcome, it is observed that an increase in the upper bound on

$$\limsup_{x \to \infty} \max_{0 \le t \le T} \frac{f(t, x)}{x}$$

or

$$\limsup_{x \to \infty} \max_{0 \le t \le T} \frac{f(t, x)}{a(t)x}$$

decreases the range on λ and vice-versa.

As a conclusion of Chap. 2, the results obtained are then applied to the mathematical models (1.9)–(1.11) with $a(t) \equiv a$, $b(t) \equiv b$, $\gamma(t) \equiv \gamma$, and $\tau(t) \equiv \tau$ being positive constants. Some explicit intervals on b are given so that the models (1.9)–(1.11) will have at least three positive T-periodic solutions. In the sequel, a new sufficient condition for the existence of at least three positive T-periodic solutions of the general Hematopoiesis model

$$x'(t) = -ax(t) + b\frac{x^m(t - \tau)}{1 + x^n(t - \tau)} \qquad (1.27)$$

is also obtained.

Chapter 3 is concerned with the existence of at least three positive T-periodic solutions of the system of differential equation (1.15). Some simple and easily verifiable sufficient conditions have been obtained for the existence of at least three positive T-periodic solutions of (1.15) such that the results can be applied to some scalar functional differential equations. As a consequence, we obtain a new sufficient condition for the existence of at least three positive T-periodic solutions of the general Hematopoiesis model with variable coefficients

$$x'(t) = -a(t)x(t) + b(t)\frac{x^m(t - \tau)}{1 + x^n(t - \tau)} \qquad (1.28)$$

and the Nicholson's Blowflies model

$$x'(t) = -a(t)x(t) + b(t)x(t - \tau)e^{-\gamma(t)x(t-\tau)}. \qquad (1.29)$$

It can be observed from the sufficient conditions assumed in Chaps. 2 and 3, that the function f needs to be unimodal, that is, the function f first increases and then it eventually decreases. This is because of the choice of the constant c_4 needed to apply the Leggett-Williams Theorem (Theorem 1.2.2). This choice of functions exclude many important class of growth functions arising in various mathematical models, such as in (1.12)–(1.14). This motivates us to study the existence of at least two positive periodic solutions of the functional differential equations (1.6) and (1.8) without using the constant c_4. Theorem 1.2.1 is used to prove the results. This is studied in Chap. 4 of this work. The results are then applied to the models (1.12)–(1.14) to show the existence of at least two positive periodic solution. Furthermore, we provide results on the existence of positive periodic solutions of Eq. (1.25) under reasonable assumptions on the functions and we apply these existence results to Eq. (1.23) to obtain information on existence of positive T-periodic solutions. An application to the existence of two positive T-periodic solutions of the model with Allee effect (1.20) is given with examples in the last section of Chap. 4.

Chapter 5 is concerned with the existence and global attractivity of the models (1.9), (1.10), (1.29), and the fishing model

$$x'(t) = -a(t)x(t) + \frac{b(t)x(t)}{1 + \left(\frac{x(t)}{p(t)}\right)^{\gamma}}, \tag{1.30}$$

where a, b, and p are T-periodic positive continuous functions and $\gamma > 1$ is a real constant. A suitable Lyapunov functional is used to study the global attractivity of equilibrium points of (1.10) and (1.29). On the other hand, the mean value theorem is used to discuss the global attractivity of an equilibrium point of (1.30). The definition of global attractivity is as follows.

Definition 1.3.1 Suppose that $x(t)$ and $\overline{x}(t)$ are two positive solutions on $[t - \tau, \infty)$, where $\tau = \max\limits_{0 \le t \le T} \tau(t)$. The solution $\overline{x}(t)$ is said to be asymptotically attractive to $x(t)$ if

$$\lim_{t \to \infty} [x(t) - \overline{x}(t)] = 0.$$

Furthermore, $\overline{x}(t)$ is called globally attractive if $\overline{x}(t)$ is asymptotically attractive to all positive solutions.

References

1. Allee, W.C.: Animal Aggregations: A Case Study in General Sociology. Chicago University Press, Chicago (1931)
2. Bai, D., Xu, Y.: Periodic solutions of first order functional differential equations with periodic deviations. Comput. Math. Appl. **53**, 1361–1366 (2007)
3. Barnett, A.: Safety in numbers. New Sci. **169**, 38–41 (2001)

4. Begon, M., Harper, J.L., Townsend, C.R.: Ecology, Individuals, Populations and Communities, 3rd edn. Blackwell Science, Oxford (1996)
5. Berec, L., Angulo, E., Courchamp, F.: Multiple Allee effects and population management. TREE **22**, 185–191 (2006)
6. Berezansky, L., Braverman, E.: Mackey-glass equation with variable coefficients. Comput. Math. Appl. **51**, 1–16 (2006)
7. Berezansky, L., Idels, L.: Stability of a time-varying fishing model with delay. Appl. Math. Lett. **21**, 447–452 (2008)
8. Boukal, D.S., Berec, L.: Single-species models of the Allee effect: extinction boundaries, sex ratios and mate encounters. J. Theor. Biol. **218**, 375–394 (2002)
9. Brockett, B.F.T., Hassall, M.: The existence of an allee effect in populations of porcellio scaber (isopoda: oniscidea). Eur. J. Soil Biol. **41**, 123–127 (2005)
10. Burgman, M.A., Ferson, S., Akcakaya, H.R.: Risk Assessment in Conservation Biology. Chapman and Hall, London (1993)
11. Castilho, C., Srinivasu, P.D.N.: Bio-economics of a renewable resource in a seasonally varying environment. Math. Biosci. **205**, 1–18 (2007)
12. Cheng, S.S., Zhang, G.: Existence of positive periodic solutions for non-autonomous functional differential equations. Electron. J. Differ. Equ. **2001**(59), 1–8 (2001)
13. Clark, C.W.: Mathematical Bioeconomics: The Optimal Management of Renewable Resources, 2nd edn. Wiley, New York (1990)
14. Courchamp, F., Clutton-Brock, T., Grenfell, B.: Inverse density dependence and the Allee effect. TREE **14**, 405–410 (1999)
15. Deimling, K.: Nonlinear Functional Analysis. Springer, Berlin (1985)
16. Driver, R.D.: Ordinary and Delay Differential Equations. Springer, New York (1976)
17. Èlsgolís, L.E., Norkin, S.B.: Introduction to the Theory and Application of Differential Equations with Deviating Arguments, Mathematics in Science and Engineering, Vol. 105. Academic Press, New York–London (1973)
18. Fowler, C.W., Baker, J.D.: A review of animal populations dynamics at extremely reduced population levels. Rep. Int. Whal. Comm. **41**, 545–554 (1991)
19. Fretwell, S.D.: Populations in a Seasonal Environment. Princeton University Press, Princeton (1972)
20. Gardner, J.L.: Winter flocking behaviour of speckled warblers and the Allee effect. Biol. Conserv. **118**, 195–204 (2004)
21. Gopalsamy, K., Weng, P.X.: Global attractivity and level crossings in a model of hematopoiesis. Bull. Inst. Math. Acad. Sinica **22**, 341–360 (1994)
22. Gopalsamy, K., Trofimchuk, S.I., Banstur, N.R.: A note on global attractivity in model of hematopoiesis. Ukr. Math. J. **50**, 3–12 (1998)
23. Graef, J.R., Qian, C., Spikes, P.W.: Oscillation and global attractivity in a periodic delay equation. Can. Math. Bull. **38**, 275–283 (1996)
24. Gurney, W.S.C., Blathe, S.P., Nishet, R.M.: Nicholson's blowflies revisited. Nature **287**, 17–21 (1980)
25. Györi, I., Ladas, G.: Oscillation Theory for Delay Differential Equations with Applications. Clarendon Press, Oxford (1991)
26. Hale, J.K.: Theory of Functional Differential Equations. Springer, New York (1977)
27. Hale, J.K., Verduyn Lunel, S.M.: Introduction to Functional Differential Equations. Springer, New York (1993)
28. Hilker, F., Langlais, M., Petrovskii, S.V., Malchow, H.: A diffusive SI model with Allee effect and application to FIV. Math. Biosci. **206**(1), 61–80 (2007)
29. Hurford, A., Hebblewhite, M., Lewis, M.A.: A spatially explicit model for an Allee effect: why wolves recolonize so slowly in greater yellowstone. Theor. Popul. Biol. **70**, 244–254 (2006)
30. Jiang, D., Wei, J.J.: Existence of positive periodic solutions for non-autonomous functional differential equations with delay (in chinese). Chinese Ann. Math. **20A**, 715–720 (1999)
31. Joseph, W., So, H., Yu, J.: Global attractivity and uniformly persistence in Nicholson's blowflies. Differ. Equ. Dyn. Syst. **2**, 11–18 (1994)

32. Kot, M.: Elements of Mathematical Ecology. Cambridge University Press, Cambridge (2001)
33. Krasnosel'skii, M.A.: Positive Solution of Operator Equations. Noordhoff, Groningen (1964)
34. Kuang, Y.: Delay Differential Equations with Applications in Population Dynamics. Academic Press, New York (1993)
35. Kulenovic, M.R.S., Ladas, G., Sficas, Y.G.: Global attractivity in Nicholson's blowflies. Appl. Anal. **43**, 109–124 (1992)
36. Kussaari, M., Saccheri, I., Camara, M., Hanski, I.: Allee effect and population dynamics in the glanville fritillary butterfly. Oikos **82**, 384–392 (1998)
37. Leggett, R.W., Williams, L.R.: Multiple positive fixed points of nonlinear operators on ordered banach spaces. Indiana Univ. Math. J. **28**, 673–688 (1979)
38. Lio, J., Yu, J.: Global asymptotic stability of nonautonomous mathematical ecological equations with distributed deviating arguments (in chinese). Acta Math. Sinica **41**, 1273–1282 (1998)
39. Mallet-Paret, J., Nussubaum, R.: Global continuation and asymptotic behaviour for periodic solutions of a differential-delay equation. Ann. Mat. Pure Appl. **145**(4), 33–128 (1986)
40. May, R.M.: Stability and Complexity in Model Ecosystems. Princeton University Press, Princeton (1973)
41. Murray, J.D.: Mathematical Biology I: An Introduction. Springer, Berlin (2004). (First Indian Reprint)
42. Odum, E.P.: Fundamentals of Ecology. Saunders, Philadelphia (1959)
43. Penteriani, V., Otalora, F., Ferrer, M.: Floater mortality within settlement areas can explain the Allee effect in breeding populations. Ecol. Model. **213**, 98–104 (2007)
44. Qian, C.: Global attractivity in a nonlinear delay differential equation with applications. Nonlinear Anal. **71**, 1893–1900 (2009)
45. Saether, B.E., Ringsby, T.H., Roskaft, E.: Life history variation, population processes and priorities in species conservation: towards a reunion of research paradigms. Oikos **77**, 217–226 (1996)
46. Saker, S.H., Agarwal, S.: Oscillation and global attractivity in a periodic Nicholson's blowflies model. Math. Comput. Model. **35**, 719–731 (2002)
47. Saker, S.H.: Oscillation and global attractivity in hematopoiesis model with periodic coefficients. Appl. Math. Comput. **142**, 477–494 (2003)
48. Saker, S.H.: Oscillation and global attractivity in hematopoiesis model with delay time. Appl. Math. Comput. **136**, 241–250 (2003)
49. Stephens, P.A., Sutherland, W.J., Freckleton, R.P.: What is the Allee effect? Oikos **87**, 185–190 (1999)
50. Tang, B., Kuang, Y.: Existence, uniqueness and asymptotic stability of periodic solutions of periodic functional differential systems. Tohoku Math. J. **49**, 217–239 (1997)
51. Wan, A., Jiang, D.: Existence of positive periodic solutions for functional differential equations. Kyushu J. Math. **56**, 193–202 (2002)
52. Wan, A., Jiang, D., Xu, X.: A new existence theory for positive periodic solutions to functional differential equations. Comput. Math. Appl. **47**, 1257–1262 (2004)
53. Wang, H.: Positive periodic solutions of functional differential equations. J. Differ. Equ. **202**, 354–366 (2004)
54. Weng, P., Liang, M.: The existence and behaviour of periodic solutions of hematopoiesis model. Math. Appl. **8**, 434–439 (1995)
55. Wu, X.M., Li, J.W., Zhou, H.Q.: A necessary and sufficient condition for the existence of positive periodic solutions of a model of hematopoiesis. Comput. Math. Appl. **54**, 840–849 (2007)
56. Wu, Y.: Existence of positive periodic solutions for a functional differential equation with a parameter. Nonlinear Anal. **68**, 1954–1962 (2008)
57. Ye, D., Fan, M., Wang, H.: Periodic solutions for scalar functional differential equations. Nonlinear Anal. **62**, 1157–1181 (2005)
58. Zhang, G., Cheng, S.S.: Positive periodic solutions of nonautonomous functional differential equations depending on a parameter. Abstr. Appl. Anal. **7**, 279–286 (2002)
59. Zhang, W., Zhu, D., Bi, P.: Existence of periodic solutions of a scalar functional differential equation via a fixed point theorem. Math. Comput. Model. **46**, 718–729 (2007)

Chapter 2
Positive Periodic Solutions of Nonlinear Functional Differential Equations with a Parameter λ

In this chapter[1], we provide several different sets of sufficient conditions for the existence of at least three positive periodic solutions to first-order functional differential equations. We will begin by considering the equation

$$x'(t) = -a(t)x(t) + \lambda f(t, x(h(t))) \tag{2.1}$$

in Sect. 2.1. Here, we assume that $R = (-\infty, \infty)$, $R_+ = [0, \infty)$, $T > 0$ is a constant, $\lambda > 0$ is a parameter, $h \in C(R, R)$, $a \in C(R, R_+)$, $a(t) \neq 0$, $a(t + T) = a(t)$, and $f \in C(R \times R_+, R_+)$ is periodic with respect to the first variable with period T.

Let X be a Banach space consisting of all positive T-periodic functions equipped with the sup norm and let K be a positive cone in X. Assume that for any $M > 0$ and $\epsilon > 0$, there exists $\delta > 0$ such that for $u, v \in K$ with $\|u\| \leq M$, $\|v\| \leq M$, and $\|u - v\| < \delta$, we have

$$\|f(t, u) - f(t, v)\| < \epsilon \tag{2.2}$$

uniformly in t.

In Sect. 2.2, we establish the existence of three positive periodic solutions of the differential equation

$$x'(t) = a(t)x(t) - \lambda f(t, x(h(t))) \tag{2.3}$$

by changing the bounds on the Green's kernel. In Sect. 2.3, we present sufficient conditions for the existence of at least three positive periodic solutions of the functional differential equation

$$x'(t) = a(t)x(t) - \lambda b(t) f(t, x(h(t))), \tag{2.4}$$

where λ, a, and f are as above, $b \in C(R, R_+)$ is T-periodic, and $\int_0^T b(t) \, dt > 0$.

---

[1] Some of the results in this chapter are based on papers [5–8].

S. Padhi et al., *Periodic Solutions of First-Order Functional Differential Equations in Population Dynamics*, DOI: 10.1007/978-81-322-1895-1_2, © Springer India 2014

If, in particular, $h(t) = t - \tau(t)$, $\tau \in C(R, R_+)$, $0 \leq \tau(t) \leq t$, then (2.1), (2.3) and (2.4) take the forms

$$x'(t) = -a(t)x(t) + \lambda f(t, x(t - \tau(t))), \qquad (2.5)$$
$$x'(t) = a(t)x(t) - \lambda f(t, x(t - \tau(t))), \qquad (2.6)$$

and

$$x'(t) = a(t)x(t) - \lambda b(t) f(t, x(t - \tau(t))), \qquad (2.7)$$

respectively.

The results obtained in this chapter can be extended to equations with multiple delays such as

$$x'(t) = -a(t)x(t) + \lambda f(t, x(t - \tau_1(t)), \ldots, x(t - \tau_m(t))) \qquad (2.8)$$

and

$$x'(t) = a(t)x(t) - \lambda f(t, x(t - \tau_1(t)), \ldots, x(t - \tau_m(t))), \qquad (2.9)$$

where $0 \leq \tau_i(t) \leq t$, $\tau_i(t + T) = \tau_i(t)$, $i = 0, 1, \ldots, m$, $f \in C(R \times R_+^m, R_+)$, and $f(t + T, x_1, x_2, \ldots, x_m) = f(t, x_1, x_2, \ldots, x_m)$.

Functional differential equations of the form (2.5) include many mathematical, ecological, and population models (either directly or after a transformation). For example:

(i) Lasota-Wazewska model

$$x'(t) = -a(t)x(t) + b(t)e^{-\gamma(t)x(t-\tau(t))}; \qquad (2.10)$$

(ii) Nicholson's blowflies model

$$x'(t) = -a(t)x(t) + b(t)x^m(t - \tau(t))e^{-\gamma(t)x^n(t-\tau(t))}; \qquad (2.11)$$

(iii) Model for red blood cell production

$$x'(t) = -a(t)x(t) + b(t)\frac{x^m(t - \tau(t))}{1 + x^n(t - \tau(t))}. \qquad (2.12)$$

Note that Eqs. (2.11) and (2.12) include (1.11) and (1.10) as special cases.

In this chapter, we prove some results on the existence of at least three positive periodic solutions to (2.1), (2.3), and (2.4) by using the Leggett-Williams multiple fixed point theorem (Theorem 1.2.2). We then apply our results to obtain some new criteria for the existence of at least three positive periodic solutions to the models (2.10)–(2.12).

Some explicit intervals on the parameter λ for the existence of solutions are given. We should also point out that the interval on λ changes according to changes in upper bound on f, that is, an increase in the upper bound on f decreases the range on λ. and vice-versa.

To study the existence of periodic solutions, we transform the given equation into an equivalent integral operator. This means that the existence of a positive periodic solution of the differential equation is equivalent to the existence of a fixed point of the operator (see Lemma 2.1.2).

The following notations are used in this chapter:

$$f^h = \limsup_{x \to h} \max_{0 \le t \le T} \frac{f(t, x)}{x}$$

and

$$\tilde{f}^h = \limsup_{x \to h} \max_{0 \le t \le T} \frac{f(t, x)}{a(t)x}.$$

2.1 Positive Periodic Solutions of the Equation $x'(t) = -a(t)x(t) + \lambda f(t, x(h(t)))$

In this section, we present sufficient conditions for the existence of three positive periodic solutions to Eq. (2.1) by using the Leggett-Williams fixed point theorem (Theorem 1.2.2 above). First we observe that Eq. (2.1) is equivalent to

$$x(t) = \lambda \int_t^{t+T} G(t, s) f(s, x(h(s))) \, ds, \qquad (2.13)$$

where

$$G(t, s) = \frac{e^{\int_t^s a(\theta) \, d\theta}}{e^{\int_0^T a(\theta) \, d\theta} - 1}$$

is the Green's function which satisfies the property

$$0 < \alpha = \frac{1}{\delta - 1} \le G(t, s) \le \frac{\delta}{\delta - 1} = \beta, \quad \text{for all} \quad s \in [t, t + T],$$

and $\delta = e^{\int_0^T a(\theta) \, d\theta}$. Since $a(t) > 0$ for $t \in [0, T]$, we have $\delta > 1$.
Let
$$X = \{x(t) : x(t) \in C(R, R), \ x(t) = x(t + T)\} \qquad (2.14)$$

with $\|x\| = \sup_{t \in [0, T]} |x(t)|$. Then X is a Banach space with the norm $\| \cdot \|$. We define a cone K in X by

$$K = \left\{ x(t) \in X : x(t) \geq \frac{\|x\|}{\delta}, \quad t \in [0, T] \right\} \tag{2.15}$$

and an operator A_λ on X by

$$(A_\lambda x)(t) = \lambda \int_t^{t+T} G(t, s) f(s, x(h(s))) \, ds. \tag{2.16}$$

Lemma 2.1.1 *The operator A_λ satisfies $A_\lambda(K) \subset K$ and $A_\lambda : K \to K$ is compact and completely continuous.*

Proof To show $A_\lambda x \in K$, we see that

$$
\begin{aligned}
(A_\lambda x)(t + T) &= \lambda \int_{t+T}^{t+2T} G(t + T, s) f(s, x(h(s))) \, ds \\
&= \lambda \int_t^{t+T} G(t + T, z + T) f(z + T, x(h(z + T))) \, dz \\
&= \lambda \int_t^{t+T} G(t, z) f(z, x(h(z))) \, dz \\
&= (A_\lambda x)(t).
\end{aligned}
$$

Hence, $A_\lambda x \in K$. Notice that for $x \in K$, we have

$$\|A_\lambda x\| \leq \lambda \beta \int_0^T f(s, x(h(s))) \, ds \tag{2.17}$$

and

$$(A_\lambda x)(t) \geq \lambda \alpha \int_0^T f(s, x(h(s))) \, ds.$$

The above inequalities imply

$$(A_\lambda x)(t) \geq \frac{\alpha}{\beta} \|A_\lambda x\| \geq \frac{\|A_\lambda x\|}{\delta}.$$

This shows that $A_\lambda x \in K$, that is, $A_\lambda(K) \subset K$.

Next, we show that A_λ is completely continuous. From assumption (2.2) on f, if $x, y \in K$ with $\|x\| \leq M$, $\|y\| \leq M$, and $\|x - y\| < \delta$, then

$$\sup_{0 \le s \le T} |f(s, u(h(s))) - f(s, v(h(s)))| < \frac{\epsilon}{\lambda \beta T}.$$

Thus,

$$|(A_\lambda x)(t) - (A_\lambda y)(t)| \le \lambda \int_t^{t+T} |G(t, s)||f(s, x(h(s))) - f(s, y(h(s)))| \, ds$$

$$\le \lambda \beta \int_0^T |f(s, x(h(s))) - f(s, y(h(s)))| \, ds$$

$$< \lambda \beta \frac{\epsilon}{\lambda \beta T} T < \epsilon.$$

Hence, A_λ is continuous. Next, we show that f maps bounded sets into bounded sets. Let $\epsilon = 1$. Then again from (2.2), we have

$$|f(s, x(h(s))) - f(s, y(h(s)))| < 1.$$

Choose a positive integer N so that $\frac{M}{N} < \delta$. Let $x \in X$ and define $x_l(t) = x(t)(\frac{l}{N})$ for $l = 0, 1, \ldots, N$. If $\|x\| \le M$, then

$$\|x_l - x_{l-1}\| = \sup_{t \in R} \left| x(t) \frac{l}{N} - x(t) \frac{l-1}{N} \right| = \frac{\|x\|}{N} \le \frac{M}{N} < \delta.$$

Hence, $|f(s, x_l(h(s))) - f(s, x_{l-1}(h(s)))| < 1$ for $s \in [0, T]$, and this gives

$$|f(s, x(h(s)))| \le \sum_{l=1}^N |f(s, x_l(h(s))) - f(s, x_{l-1}(h(s)))| + |f(s, 0)|$$

$$\le N + \|f(s, 0)\| =: M_1.$$

Now, from (2.17) and the fact that $t \in [0, T]$, we have

$$\|A_\lambda x\| \le \lambda \beta \int_0^T f(s, x(h(s))) \, ds \le \lambda \beta T M_1. \tag{2.18}$$

Finally, we have

$$\frac{d}{dt}(A_\lambda x)(t) = \lambda[G(t, t+T) f(t+T, x(h(t+T))) - G(t, t) f(t, x(h(t)))]$$

$$- a(t)(A_\lambda x)(t).$$

From (2.18) and the choice of M_1, we obtain

$$\left|\frac{d}{dt}(A_\lambda x)(t)\right| \leq \lambda\beta M_1[\|a\|T + 2].$$

Hence, $\{(A_\lambda x) : x \in K, \|x\| < M\}$ is a family of uniformly bounded and equicontinuous function on $[0, T]$. Thus the operator A_λ is completely continuous by the Ascoli-Arzelà theorem (see Royden [9, p. 169]). This completes the proof of the Lemma.   □

Lemma 2.1.2 *The existence of a positive periodic solution of (2.1) is equivalent to the existence of a fixed point of the operator A_λ in K.*

Proof First, for $x \in K$ and $A_\lambda x = x$, we have

$$\begin{aligned}
x'(t) &= \frac{d}{dt}\left(\lambda\int_t^{t+T} G(t, s)f(s, x(h(s)))\,ds\right) \\
&= \lambda G(t, t+T)f(t+T, x(h(t+T))) - \lambda G(t, t)f(t, x(h(t))) \\
&\quad + \lambda\int_t^{t+T} \frac{\partial}{\partial t}G(t, s)f(s, x(h(s)))\,ds \\
&= \lambda[G(t, t+T) - G(t, t)]f(t, x(h(t))) - a(t)(A_\lambda x)(t) \\
&= \lambda f(t, x(h(t))) - a(t)x(t),
\end{aligned}$$

since $\frac{\partial}{\partial t}G(t, s) = -a(t)G(t, s)$.

Next, if x is a positive T-periodic solution, we have

$$\begin{aligned}
(A_\lambda x)(t) &= \lambda\int_t^{t+T} G(t, s)f(s, x(h(s)))\,ds \\
&= \int_t^{t+T} G(t, s)(x'(s) + a(s)x(s))\,ds \\
&= \int_t^{t+T} G(t, s)x'(s)\,ds + \int_t^{t+T} G(t, s)a(s)x(s)\,ds \\
&= [G(t, s)x(s)]_t^{t+T} - \int_t^{t+T}\left(\frac{\partial}{\partial s}G(t, s)\right)x(s)\,ds + \int_t^{t+T} G(t, s)a(s)x(s)\,ds
\end{aligned}$$

$$= [G(t, t+T) - G(t, t)]x(t) - \int_t^{t+T} G(t, s)a(s)x(s)\,ds$$

$$+ \int_t^{t+T} G(t, s)a(s)x(s)\,ds$$

$$= x(t),$$

since $\frac{\partial}{\partial s}G(t, s) = a(s)G(t, s)$. Clearly, $(A_\lambda x)(t) \geq 0$ for $t \in [0, T]$, and this completes the proof of the Lemma.   □

In order to prove the existence of a positive T-periodic solution of (2.1), in view of Lemma 2.1.2, it suffices to show that the operator in (2.16) has a fixed point.

Theorem 2.1.1 *Let $f^\infty < 1$ hold. Assume that there are constants $0 < c_1 < c_2$ such that*

(H_1) *$f(t, x) \geq 2\delta c_2$ for $x \in K$, $c_2 \leq x \leq \delta c_2$, and $0 \leq t \leq T$, and*

(H_2) *$f(t, x) < c_1$ for $x \in K$, $0 \leq x \leq c_1$, and $0 \leq t \leq T$.*

Then (2.1) has at least three positive T-periodic solutions for

$$\frac{\delta - 1}{2\delta T} < \lambda < \frac{\delta - 1}{\delta T}.$$

Proof We consider the Banach space X defined in (2.14) and the cone K as in (2.15). Since $f^\infty < 1$, there exist $\epsilon \in (0, 1)$ and $\theta > 0$ such that $f(t, x) \leq \epsilon x$ for $x \geq \theta$. Let

$$\gamma = \max_{0 \leq x \leq \theta,\, 0 \leq t \leq T} f(t, x).$$

Then $f(t, x) \leq \epsilon x + \gamma$ for $x \geq 0$ and $0 \leq t \leq T$. Choose

$$c_4 > \max\left\{\frac{\gamma}{1 - \epsilon}, \delta c_2\right\}.$$

Then for $x \in \overline{K}_{c_4}$, we have

$$\|A_\lambda x\| = \sup_{0 \leq t \leq T} \lambda \int_t^{t+T} G(t, s)f(s, x(h(s)))\,ds$$

$$\leq \frac{\lambda \delta}{\delta - 1} \int_0^T f(s, x(h(s)))\,ds$$

$$\leq \frac{\lambda\delta}{\delta - 1} \int_0^T (\epsilon x(h(s)) + \gamma) \, ds$$

$$\leq \frac{\lambda\delta}{\delta - 1} \int_0^T (\epsilon \|x\| + \gamma) \, ds$$

$$\leq \frac{\lambda\delta(\epsilon c_4 + \gamma)}{\delta - 1} T$$

$$< c_4.$$

Hence $A_\lambda : \overline{K}_{c_4} \to \overline{K}_{c_4}$.

Next, define a nonnegative concave continuous functional ψ on K by $\psi(x) = \min_{t \in [0,T]} x(t)$. Let $c_3 = \delta c_2$ and $\phi_0(t) = \phi_0$, where ϕ_0 is any given number satisfying $c_2 < \phi_0 < c_3$. Then $\phi_0 \in \{x \in K(\psi, c_2, c_3) : \psi(x) > c_2\}$. For $x \in K(\psi, c_2, c_3)$, by (H_1) we have

$$\psi(A_\lambda x) = \min_{0 \leq t \leq T} \lambda \int_t^{t+T} G(t, s) f(s, x(h(s))) \, ds$$

$$\geq \frac{\lambda}{\delta - 1} \int_0^T f(s, x(h(s))) \, ds$$

$$\geq \frac{\lambda}{\delta - 1} 2 \delta c_2 T$$

$$> c_2.$$

Thus, property (i) of Theorem 1.2.2 is satisfied.

Now for $x \in \overline{K}_{c_1}$, (H_2) gives

$$\|A_\lambda x\| = \sup_{0 \leq t \leq T} \lambda \int_t^{t+T} G(t, s) f(s, x(h(s))) \, ds$$

$$\leq \frac{\lambda\delta}{\delta - 1} \int_0^T f(s, x(h(s))) \, ds$$

$$\leq \frac{\lambda\delta}{\delta - 1} c_1 T$$

$$< c_1,$$

that is, $A_\lambda x \in \overline{K}_{c_1}$. Hence, property (ii) of Theorem 1.2.2 is satisfied.

Finally, for $x \in K(\psi, c_2, c_4)$ with $\|A_\lambda x\| > c_3$, we have

$$c_3 < \|A_\lambda x\| \leq \frac{\lambda \delta}{\delta - 1} \int_0^T f(s, x(h(s))) \, ds,$$

which implies that

$$\psi(A_\lambda x) \geq \frac{\lambda}{\delta - 1} \int_0^T f(s, x(h(s))) \, ds$$

$$> \frac{c_3}{\delta}$$

$$= c_2.$$

This proves that property (iii) of Theorem 1.2.2 holds. Therefore, (2.1) has at least three positive T-periodic solutions. This completes the proof of the theorem. $\quad\square$

Theorem 2.1.2 *Let* $f^\infty < \frac{1}{\beta}$ *hold and assume that there exist* $0 < c_1 < c_2$ *such that*

(H_3) $f(t, x) \geq \delta(\delta - 1)c_2$ *for* $x \in K$, $c_2 \leq x \leq \delta c_2$, *and* $0 \leq t \leq T$,
and

(H_4) $f(t, x) < \frac{1}{\beta}c_1$ *for* $x \in K$, $0 \leq x \leq c_1$, *and* $0 \leq t \leq T$.

Then Eq. (2.1) has at least three positive T-periodic solutions for

$$\frac{1}{\delta T} < \lambda < \frac{1}{T}.$$

The proof of this theorem is essentially the same as the proof of Theorem 2.1.1, but we include it here for the sake of completeness.

Proof Consider the Banach space X given in (2.14) and the cone K as in (2.15). Since $f^\infty < \frac{1}{\beta}$, there exist $\epsilon \in (0, \frac{1}{\beta})$ and $\theta > 0$ such that $f(t, x) \leq \epsilon x$ for $x \geq \theta$ and $0 \leq t \leq T$. Suppose that

$$\gamma = \max_{0 \leq x \leq \theta, \, 0 \leq t \leq T} f(t, x).$$

Then, $f(t, x) \leq \epsilon x + \gamma$ for $x \geq 0$ and $0 \leq t \leq T$. Choosing $c_4 > \max\left\{\frac{\gamma \beta}{1 - \beta \epsilon}, \delta c_2\right\}$ and proceeding as in the proof of the Theorem 2.1.1, it can be shown that $A_\lambda : \overline{K}_{c_4} \to \overline{K}_{c_4}$.

Next, we define a nonnegative concave continuous functional ψ on K by $\psi(x) = \min_{t \in [0, T]} x(t)$. Choosing $c_3 = \delta c_2$ and $\phi_0(t) = \phi_0 \in R$ satisfying $c_2 < \phi_0 < c_3$, we see that $\phi_0 \in \{x \in K(\psi, c_2, c_3) : \psi(x) > c_2\}$ is nonempty. Then, for $x \in K(\psi, c_2, c_3)$, using (H_3) we have

$$
\begin{aligned}
\psi(A_\lambda x) &\geq \frac{\lambda}{\delta - 1} \int_0^T f(s, x(h(s)))\, ds \\
&\geq \frac{\lambda}{\delta - 1} \delta(\delta - 1)\, c_2\, T \\
&> c_2.
\end{aligned}
$$

Now for $x \in \overline{K}_{c_1}$, from (H_4),

$$
\begin{aligned}
\|A_\lambda x\| &\leq \frac{\lambda \delta}{\delta - 1} \int_0^T f(s, x(h(s)))\, ds \\
&< \frac{\lambda \delta}{\delta - 1} c_1 \frac{\delta - 1}{\delta} T \\
&< c_1.
\end{aligned}
$$

Furthermore, for $x \in K(\psi, c_2, c_4)$ with $\|A_\lambda x\| > c_3$, we have

$$
c_3 < \|A_\lambda x\| \leq \frac{\delta}{\delta - 1} \lambda \int_0^T f(s, x(h(s)))\, ds,
$$

which in turn implies

$$
\begin{aligned}
\psi(A_\lambda x) &\geq \frac{1}{\delta - 1} \lambda \int_0^T f(s, x(h(s)))\, ds \\
&> \frac{c_3}{\delta} \\
&= c_2.
\end{aligned}
$$

Therefore, by Theorem 1.2.2, Eq. (2.1) has at least three positive T-periodic solutions, and this completes the proof of the theorem.   □

Theorem 2.1.3 *Let $f^\infty < \frac{1}{\beta^2}$ and assume that there are constants $0 < c_1 < c_2$ such that*

(H_5) $f(t, x) \geq \frac{c_2}{\alpha}$ *for $x \in K$, $c_2 \leq x \leq \delta c_2$ and $0 \leq t \leq T$*

and

(H_6) $f(t, x) < \frac{c_1}{\beta^2}$ for $x \in K$, $0 \le x \le c_1$ and $0 \le t \le T$.

Then Eq. (2.1) has at least three positive T-periodic solutions for

$$\frac{1}{T} < \lambda < \frac{\beta}{T}.$$

Proof Choose the Banach space X and cone K as in (2.14) and (2.15), respectively. Clearly, $f^\infty < \frac{1}{\beta^2}$ implies that there exist $\epsilon \in (0, \frac{1}{\beta^2})$ and $\theta > 0$ such that $f(t, x) < \epsilon x$ for $x \ge \theta$ and $0 \le t \le T$. Choose γ as in the proof of Theorem 2.1.1. It is easy to show that $A_\lambda : \overline{K}_{c_4} \to \overline{K}_{c_4}$, where

$$c_4 > \max \left\{ \frac{\beta^2 \gamma}{1 - \beta^2 \epsilon}, \delta c_2 \right\}.$$

Now, define a nonnegative concave continuous functional ψ on K as in the proof of Theorem 2.1.1. Then for $x \in K(\psi, c_2, c_3)$, from (H_5) we have $\|A_\lambda x\| > c_2$. In addition, for $x \in \overline{K}_{c_1}$, (H_6) implies

$$\|A_\lambda x\| \le \beta \lambda \int_0^T f(s, x(h(s))) \, ds$$

$$< \beta \lambda \int_0^T \frac{c_1}{\beta^2} \, ds$$

$$< c_1.$$

The rest of the proof is the same as that of Theorem 2.1.1. Therefore, (2.1) has at least three positive T-periodic solutions. The theorem is now proved. $\qquad \square$

Theorem 2.1.4 *Let* $f^\infty < \frac{(\delta-1)^2}{\delta^3}$ *and assume there exist constants* $0 < c_1 < c_2$ *such that* (H_5) *holds and*

(H_7) $f(t, x) < \frac{(\delta-1)^2}{\delta^3} c_1$ for $x \in K$, $0 \le x \le c_1$ and $0 \le t \le T$.

Then Eq. (2.1) has at least three positive T-periodic solutions for

$$\frac{1}{T} < \lambda < \frac{\delta^2}{(\delta - 1)T}.$$

Proof Since $f^\infty < \frac{(\delta-1)^2}{\delta^3}$, there exist $\epsilon \in \left(0, \frac{(\delta-1)^2}{\delta^3}\right)$ and $\theta > 0$ such that $f(t, x) \le \epsilon x$ for $x \ge \theta$. Suppose that $\gamma = \max_{0 \le x \le \theta, \, 0 \le t \le T} f(t, x)$. Set

$$c_4 > \max \left\{ \frac{\delta^3 T}{(\delta - 1)^2 - \delta^3 \epsilon}, \delta c_2 \right\}.$$

Proceeding along the lines of the proof of Theorem 2.1.1, we can show that $A_\lambda : \overline{K}_{c_4} \to \overline{K}_{c_4}$.

With the nonnegative concave continuous functional ψ on K defined as in Theorem 2.1.1 and using (H_5), we have

$$\psi(A_\lambda x) \geq \frac{\lambda}{\delta - 1} \int_0^T f(s, x(h(s)))\, ds$$

$$\geq \frac{\lambda}{\delta - 1}(\delta - 1)c_2 T$$

$$> c_2.$$

Next, using (H_7), for $x \in \overline{K}_{c_1}$,

$$\|A_\lambda x\| \leq \lambda \frac{\delta}{\delta - 1} \int_0^T f(s, x(h(s)))\, ds$$

$$< \lambda \frac{\delta}{\delta - 1} \frac{(\delta - 1)^2}{\delta^3} c_1 T$$

$$< c_1.$$

The remainder of the proof is similar to the proof of Theorem 2.1.1. Consequently, (2.1) has at least three positive T-periodic solutions and this completes the proof of the theorem. □

Theorem 2.1.5 *Assume that $f^\infty < 1$, $f^0 < 1$, and there is a constant $c_2 > 0$ such that (H_1) holds. Then (2.1) has at least three positive periodic solutions for*

$$\frac{\delta - 1}{2\delta T} < \lambda < \frac{\delta - 1}{\delta T}.$$

Proof Since $f^\infty < 1$, there exist $0 < \delta_1 < 1$ and $\xi_1 > 0$ such that

$$f(t, x) \leq \delta_1 x \quad \text{for } x \geq \xi_1.$$

Let $\beta_1 = \max_{0 \leq x \leq \xi_1, 0 \leq t \leq T} f(t, x)$. Then,

$$f(t, x) \leq \delta_1 x + \beta_1 \quad \text{for } 0 \leq x < \infty.$$

Choose

$$c_4 > \max \left\{ \frac{\beta_1}{1 - \delta_1}, \delta c_2 \right\}.$$

For $x \in \overline{K}_{c_4}$, we have

$$\|A_\lambda x\| = \sup_{0 \le t \le T} \lambda \int_t^{t+T} G(t, s) f(s, x(h(s))) \, ds$$

$$\le \frac{\lambda \delta}{\delta - 1} \int_0^T f(s, x(h(s))) \, ds$$

$$\le \frac{\lambda \delta}{\delta - 1} \int_0^T (\delta_1 x(h(s)) + \beta_1) \, ds$$

$$\le \frac{\lambda \delta}{\delta - 1} \int_0^T (\delta_1 \|x\| + \beta_1) \, ds$$

$$\le \frac{\lambda \delta (\delta_1 c_4 + \beta_1)}{\delta - 1} T$$

$$< c_4.$$

This shows $A_\lambda : \overline{K}_{c_4} \to \overline{K}_{c_4}$. Now setting $c_3 = \delta c_2$, defining the nonnegative concave continuous functional $\psi(x) = \min_{t \in [0,T]} x(t)$, and using (H_1), we can show that condition (i) of Theorem 1.2.2 holds.

Since $f^0 < 1$, there exist $0 < \delta_2 < 1$ and $0 < \xi_2 < \frac{c_2}{2}$ such that

$$f(t, x) \le \delta_2 x \quad for \ 0 \le x \le \xi_2.$$

Set $0 < c_1 = \xi_2$; then for $x \in \overline{K}_{c_1}$, we have

$$\|A_\lambda x\| = \sup_{0 \le t \le T} \lambda \int_0^T G(t, s) f(s, x(h(s))) \, ds$$

$$\le \lambda \delta_2 \frac{\delta}{\delta - 1} c_1 T$$

$$\le \delta_2 c_1$$

$$< c_1.$$

Thus, condition (ii) of Theorem 1.2.2 is satisfied. The proof of (iii) in Theorem 1.2.2 is easy. Consequently, (2.1) has at least three positive T-periodic solutions, which proves the theorem. $\qquad \square$

Example 2.1.1 Consider the equation

$$x'(t) = -\frac{\log 3}{2}|\sin t|x(t) + \frac{12}{35\pi}e^9 x^3(t)e^{-x(t)}, \quad t \geq 0. \tag{2.19}$$

Here $a(t) = \frac{\log 3}{2}|\sin t|$ and $T = \pi$. Setting $\lambda = \frac{3}{7\pi}$, we see that $f(t, x) = \frac{4}{5}e^9 x^3 e^{-x}$. Clearly, $f^\infty < 1$ and $f^0 < 1$. Furthermore, $\int_0^\pi a(t)\,dt = \log 3$ implies that $\delta = 3$. Hence, $\alpha = \frac{1}{\delta-1} = \frac{1}{2}$ and $\beta = \frac{\delta}{\delta-1} = \frac{3}{2}$. Choosing $c_2 = 3$, we have $c_3 = \delta c_2 = 9$. Then it is easy to see that

$$f(t, x) = \frac{4}{5}e^9 x^3 e^{-x} > \frac{108}{5} > 18 = 2\delta c_2 \quad \text{for } c_2 \leq x \leq c_3,$$

that is, (H_1) holds. In addition, $\lambda = \frac{3}{7\pi} \in \left(\frac{1}{3\pi}, \frac{2}{3\pi}\right) = \left(\frac{\delta-1}{2\delta T}, \frac{\delta-1}{\delta T}\right)$. Thus, by Theorem 2.1.5, Eq. (2.19) has at least three positive T-periodic solutions.

Remark 2.1.1 In general, it is difficult to obtain a function $f(t, x(t))$ satisfying (H_1) and (H_2), or (H_3) and (H_4), or (H_5) and (H_6), or (H_5) and (H_7), simultaneously. From Theorem 2.1.5, it is easy to verify that the conditions (H_2), (H_4), (H_6), and (H_7) can be replaced by the conditions $f^0 < 1$, $f^0 < \frac{1}{\beta}$, $f^0 < \frac{1}{\beta^2}$ and $f^0 < \frac{(\delta-1)^2}{\delta^3}$, respectively. One may proceed along the lines of the proof of Theorem 2.1.5 to finish the proof.

Remark 2.1.2 It is not difficult to check that

$$\int_t^{t+T} a(s)G(t, s)\,ds \equiv 1.$$

This leads us to obtain the following new sufficient conditions for the existence of at least three positive T-periodic solutions of (2.1) using the symbol $\tilde{f}^h$ defined earlier.

Theorem 2.1.6 *Let $\tilde{f}^\infty < T$ hold and assume that there are constants $0 < c_1 < c_2$ such that (H_1) and (H_4) hold. Then (2.1) has at least three positive T-periodic solutions for*

$$\frac{\delta-1}{2\delta T} < \lambda < \frac{1}{T}.$$

Proof From $\tilde{f}^\infty < T$, it follows that there exist $\epsilon \in (0, T)$ and $\theta > 0$ such that $f(t, x) \leq \epsilon a(t)x$ for $x \geq \theta$ and $0 \leq t \leq T$. Let

$$\gamma = \max_{0 \leq x \leq \theta,\, 0 \leq t \leq T} f(t, x).$$

Then $f(t, x) \le \epsilon a(t)x + \gamma$ for $x \ge 0$ and $0 \le t \le T$. Choose

$$c_4 > \max \left\{ \frac{\delta \gamma T}{(\delta - 1)(T - \epsilon)}, \delta c_2 \right\}.$$

Then, for $x \in \overline{K}_{c_4}$, we have

$$\|A_\lambda x\| = \sup_{0 \le t \le T} \lambda \int_t^{t+T} G(t, s) f(s, x(h(s))) \, ds$$

$$\le \sup_{0 \le t \le T} \lambda \int_t^{t+T} G(t, s)(a(s)x(h(s)) \epsilon + \gamma) \, ds$$

$$\le \sup_{0 \le t \le T} \lambda \int_t^{t+T} G(t, s)(a(s)\|x\| \epsilon + \gamma) \, ds$$

$$\le \lambda \left[\epsilon c_4 \sup_{0 \le t \le T} \int_t^{t+T} a(s)G(t, s) \, ds + \sup_{0 \le t \le T} \gamma \int_t^{t+T} G(t, s) \, ds \right]$$

$$\le \lambda \left[\epsilon c_4 + \frac{\gamma \delta}{\delta - 1} T \right]$$

$$< \frac{1}{T} \left[\epsilon c_4 + \frac{\gamma \delta}{\delta - 1} T \right]$$

$$< c_4.$$

Hence, $A_\lambda : \overline{K}_{c_4} \to \overline{K}_{c_4}$.

Next, we define a nonnegative concave continuous functional ψ on K by $\psi(x) = \min_{t \in [0,T]} x(t)$. Then $\psi(x) \le \|x\|$. Let $c_3 = \delta c_2$ and $\phi_0(t) = \phi_0$, where ϕ_0 is any given number satisfying $c_2 < \phi_0(t) < c_3$. Then $\phi_0 \in \{x : x \in K(\psi, c_2, c_3), \psi(x) > c_2\}$. Furthermore, for $x \in K(\psi, c_2, c_3)$, from (H_1) we have

$$\psi(A_\lambda x) = \min_{0 \le t \le T} \lambda \int_t^{t+T} G(t, s) f(s, x(h(s))) \, ds$$

$$\ge \frac{1}{\delta - 1} \lambda \int_0^T f(s, x(h(s))) \, ds$$

$$\ge \frac{\lambda}{\delta - 1} 2 \delta c_2 T$$

$$> c_2.$$

Now, let $x \in \overline{K}_{c_1}$; then, using (H_4),

$$
\begin{aligned}
\|A_\lambda x\| &= \sup_{0 \le t \le T} \lambda \int_t^{t+T} G(t, s) f(s, x(h(s))) \, ds \\
&\le \frac{\lambda \delta}{\delta - 1} \int_0^T f(s, x(h(s))) \, ds \\
&\le \frac{\lambda \delta}{\delta - 1} c_1 \frac{\delta - 1}{\delta} T \\
&< c_1,
\end{aligned}
$$

that is, $A_\lambda x \in \overline{K}_{c_1}$.

Finally, for $x \in K(\psi, c_2, c_4)$ with $\|A_\lambda x\| > c_3$, we have

$$
c_3 < \|A_\lambda x\| \le \frac{\delta}{\delta - 1} \lambda \int_0^T f(s, x(h(s))) \, ds,
$$

which in turn implies that

$$
\begin{aligned}
\psi(A_\lambda x) &\ge \frac{1}{\delta - 1} \lambda \int_0^T f(s, x(h(s))) \, ds \\
&> \frac{c_3}{\delta} \\
&= c_2.
\end{aligned}
$$

Hence, all the conditions of Theorem 1.2.2 are satisfied and so Eq. (2.1) has at least three positive T-periodic solutions. This completes the proof of the theorem. □

Theorem 2.1.7 *Let $\tilde{f}^\infty < T$. Assume that there exist constants $0 < c_1 < c_2$ such that (H_4) and*

(H_8) $f(t, x) \ge 2(\delta - 1) c_2$ *for $x \in K$, $c_2 \le x \le \delta c_2$ and $0 \le t \le T$.*

Then (2.1) has at least three positive T-periodic solutions for

$$
\frac{1}{2T} < \lambda < \frac{1}{T}.
$$

Proof The proof of the theorem is quite similar to proof of Theorem 2.1.6. Here, we use (H_8) in place of (H_1) in the following way to show condition (i) of Theorem 1.2.2. Define ψ on K by $\psi(x) = \min_{t \in [0,T]} x(t)$ and set $c_3 = \delta c_2$. Then,

$$\psi(A_\lambda x) = \min_{0 \le t \le T} \lambda \int_t^{t+T} G(t, s) f(s, x(h(s))) \, ds$$

$$\ge \frac{1}{\delta - 1} \lambda \int_0^T f(s, x(h(s))) \, ds$$

$$\ge \frac{\lambda}{\delta - 1} 2 (\delta - 1) c_2 T$$

$$> c_2.$$

Thus, by Theorem 1.2.2, (2.1) has at least three positive T-periodic solutions.   □

Theorem 2.1.8 *Let $\tilde{f}^\infty < T$ and $\tilde{f}^0 < T$. In addition, assume that there exists $c_2 > 0$ such that (H_1) holds. Then there exist at least three positive T-periodic solutions of (2.1) for*

$$\frac{\delta - 1}{2\delta T} < \lambda < \frac{1}{T}.$$

Proof Since $\tilde{f}^\infty < T$, there exist $0 < \delta_1 < T$ and $\xi_1 > 0$ such that

$$f(t, x) \le \delta_1 a(t) x \quad \text{for } x \ge \xi_1 \text{ and } 0 \le t \le T.$$

Let

$$\gamma = \max_{0 \le x \le \xi_1, 0 \le t \le T} f(t, x).$$

Then $f(t, x) \le \delta_1 a(t) x + \gamma$ for $x \ge 0$ and $0 \le t \le T$, so choosing c_4 as in the proof of Theorem 2.1.6, we can show that $A_\lambda : \overline{K}_{c_4} \to \overline{K}_{c_4}$. Defining a concave continuous functional ψ on K by $\psi(x) = \min_{t \in [0,T]} x(t)$ and using (H_1), we can prove that the condition (i) of Theorem 1.2.2 holds.

Now $\tilde{f}^0 < T$ implies there exist δ_2, $0 < \delta_2 < T$, and $\frac{c_2}{2} > \xi_2 > 0$ such that

$$f(t, x) \le \delta_2 a(t) x \quad \text{for } 0 \le x \le \xi_2 \text{ and } 0 \le t \le T.$$

Set $0 < c_1 = \xi_2$; then $0 < c_1 < c_2$. For $x \in \overline{K}_{c_1}$, we have

$$\|A_\lambda x\| = \sup_{0 \le t \le T} \lambda \int_t^{t+T} G(t, s) f(s, x(h(s))) \, ds$$

$$\le \lambda \delta_2 \sup_{0 \le t \le T} \int_0^T G(t, s) a(s) \|x\| \, ds$$

$$\leq \lambda c_1 \delta_2 \sup_{0 \leq t \leq T} \int_0^T a(s) G(t, s) \, ds$$

$$\leq \lambda c_1 \delta_2$$

$$< c_1,$$

that is, the condition (ii) of Theorem 1.2.2 is satisfied. In a similar way to what was done in the proof of Theorem 2.1.6, we can show that condition (iii) of Theorem 1.2.2 is satisfied. Hence, there exist at least three positive T-periodic solutions of (2.1) proving the theorem.   □

Theorem 2.1.9 *Let $\tilde{f}^\infty < T$, $\tilde{f}^0 < T$ and (H_8) hold. Then there exist at least three positive T-periodic solutions of (2.1) for*

$$\frac{1}{2T} < \lambda < \frac{1}{T}.$$

Proof Since $\tilde{f}^\infty < T$ and $\tilde{f}^0 < T$, we can proceed as in the proof of Theorem 2.1.8 to prove that $A_\lambda : \overline{K}_{c_4} \to \overline{K}_{c_4}$, and conditions (ii) and (iii) of Theorem 1.2.2 hold. To complete the proof of the theorem, it remains to show the condition (i) of Theorem 1.2.2 is satisfied. We consider the nonnegative concave continuous functional ψ as before. Then, for $x \in K(\psi, c_2, c_3)$, we have

$$\psi(A_\lambda x) \geq \frac{\lambda}{\delta - 1} \int_0^T f(s, x(h(s))) \, ds$$

$$\geq \frac{\lambda}{\delta - 1} 2(\delta - 1) c_2 T$$

$$> c_2$$

by (H_8). Hence, condition (i) of Theorem 1.2.2 is satisfied, and this completes the proof of the theorem.   □

In [10], Zhang et al. proved a theorem for the existence of at least two positive T-periodic solutions of (2.8) (see [10, Theorem 3.2]). Applying this theorem to (2.5), we obtain the following result.

Theorem 2.1.10 *Let $\lambda = 1$, $\tilde{f}^0 < 1$, and $\tilde{f}^\infty < 1$. In addition, assume that there exists $\rho > 0$ such that $f(t, x) > a(t)|x|$ for $\mu\rho < |x| < \rho$, where $\mu = exp\{-\int_0^T a(s) \, ds\}$. Then (2.5) has at least two positive T-periodic solutions x_1 and x_2 such that*

$$0 < \|x_1\| < \rho < \|x_2\|.$$

The following Corollary 2.1.1 follows from Theorems 2.1.8 and 2.1.9.

Corollary 2.1.1 *Let $\tilde{f}^0 < T$ and $\tilde{f}^\infty < T$, and assume that there exists a constant $c_2 > 0$ such that either (H_1) or (H_8) holds. Then (2.5) has at least three positive T-periodic solutions for*

$$\frac{\delta - 1}{2\,\delta T} < \lambda < \frac{1}{T}.$$

Corollary 2.1.1 is different from Theorem 2.1.10. Indeed, the upper bound on $\tilde{f}^0$ and $\tilde{f}^\infty$ considered in Corollary 2.1.1 is the general period T, whereas in Theorem 2.1.10 it is 1. However, a range on λ has been given in Corollary 2.1.1.

Theorem 2.1.11 *Let $\tilde{f}^\infty < T$. Assume that there are constants $0 < c_1 < c_2$ such that (H_1) holds and*

(H_9) $f(t, x) < x$ for $0 \leq x \leq c_1$ and $0 \leq t \leq T$.

Then there exist at least three positive T-periodic solution of (2.1) for

$$\frac{\delta - 1}{2\delta T} < \lambda < \frac{\delta - 1}{\delta T}.$$

Proof Since $\tilde{f}^\infty < T$, there exist $\epsilon \in (0, T)$ and $\theta > 0$ such that $f(t, x) \leq \epsilon a(t)x$ for $x \geq \theta$ and $0 \leq t \leq T$. Let $\gamma = \max\limits_{0 \leq x \leq \theta, 0 \leq t \leq T} f(t, x)$. Then $f(t, x) \leq \epsilon a(t)x + \gamma$ for $x \geq 0$ and $0 \leq t \leq T$. Now choosing

$$c_4 > \max \left\{ \frac{\delta \gamma T}{\delta(T - \epsilon) + \epsilon}, \delta c_2 \right\},$$

we can prove that $A_\lambda : \overline{K}_{c_4} \to \overline{K}_{c_4}$. However, we need the following argument to prove that condition (ii) of Theorem 1.2.2 holds. From (H_9) we have, for $x \in \overline{K}_{c_1}$

$$\|A_\lambda x\| = \sup_{0 \leq t \leq T} \lambda \int_t^{t+T} G(t, s) f(s, x(h(s))) \, ds$$

$$\leq \lambda \frac{\delta}{\delta - 1} c_1 T$$

$$< c_1.$$

The proof of the condition (iii) of Theorem 1.2.2 is easy and hence is omitted. Thus, (2.1) has at least three positive T-periodic solutions proving the theorem. $\qquad \square$

2.2 Positive Periodic Solutions of the Equation
$x'(t) = a(t)x(t) - \lambda f(t, x(h(t)))$

This section is concerned with the existence of at least three positive periodic solutions of Eq. (2.3). In this case, Eq. (2.3) is equivalent to the integral equation (2.13) with a different kernel, namely,

$$G(t, s) = \frac{e^{-\int_t^s a(\theta)\,d\theta}}{1 - e^{-\int_0^T a(\theta)\,d\theta}}, \qquad (2.20)$$

which has the property

$$0 < \frac{\delta}{1 - \delta} \le G(t, s) \le \frac{1}{1 - \delta} \quad \text{for } s \in [t, t + T], \qquad (2.21)$$

where $\delta = e^{-\int_0^T a(\theta)\,d\theta} < 1$.

We consider the Banach space X as in (2.14) and a cone K in X given by

$$K = \{x \in X : x(t) \ge \delta\|x\|, \, 0 \le t \le T\}. \qquad (2.22)$$

Defining the operator A_λ by (2.16), we see that $A_\lambda(K) \subset K$. As in Sect. 2.1, it can be proved that the existence of a positive periodic solution of (2.3) is equivalent to the existence of a fixed point of A_λ in the above cone and that $A_\lambda : \overline{K}_{c_4} \to \overline{K}_{c_4}$ is completely continuous.

Theorem 2.2.1 *Assume that $\tilde{f}^\infty < T$ and there exist $0 < c_1 < c_2$ such that*

(H_{10}) $f(t, x) \ge \frac{c_2}{\delta}$ *for $x \in K$, $c_2 \le x \le \frac{c_2}{\delta}$ and $0 \le t \le T$*

and

(H_{11}) $f(t, x) < (1 - \delta)c_1$ *for $x \in K$, $0 \le x \le c_1$ and $0 \le t \le T$.*

Then Eq. (2.3) has at least three positive T-periodic solutions for

$$\frac{1 - \delta}{T} < \lambda < \frac{1}{T}.$$

Proof Since $\tilde{f}^\infty < T$, there exist $\delta_1 \in (0, T)$ and $\mu > 0$ such that $f(t, x) \le \delta_1 a(t)x$ for $x \ge \mu$ and $0 \le t \le T$. Let $M = \max\limits_{0 \le x \le \mu, \, 0 \le t \le T} f(t, x)$. Then $f(t, x) \le \delta_1 a(t)x + M$ for $x \ge 0$ and $0 \le t \le T$. Choose

$$c_4 > \max\left\{\frac{MT}{(1-\delta)(T-\delta_1)}, \frac{c_2}{\delta}\right\}.$$

Now for $x \in \overline{K}_{c_4}$, we have

$$\|A_\lambda x\| = \sup_{0 \le t \le T} \lambda \int_t^{t+T} G(t,s) f(s, x(h(s)))\, ds$$

$$\le \sup_{0 \le t \le T} \lambda \int_t^{t+T} G(t,s)(a(s)x(h(s))\delta_1 + M)\, ds$$

$$\le \sup_{0 \le t \le T} \lambda \int_t^{t+T} G(t,s)(a(s)\|x\|\delta_1 + M)\, ds$$

$$\le \lambda \left[\delta_1 c_4 \sup_{0 \le t \le T} \int_t^{t+T} a(s)G(t,s)\, ds + \sup_{0 \le t \le T} M \int_t^{t+T} G(t,s)\, ds \right]$$

$$\le \lambda \left[\delta_1 c_4 + \frac{M}{1-\delta} T \right]$$

$$< c_4.$$

Hence, $A_\lambda : \overline{K}_{c_4} \to \overline{K}_{c_4}$.

Next, we define a nonnegative concave continuous functional ψ on K by $\psi(x) = \min_{t \in [0,T]} x(t)$. Let $c_3 = \frac{c_2}{\delta}$ and $\phi_0(t) = \phi_0$ satisfy $c_2 < \phi_0 < c_3$. This shows $\phi_0 \in \{x : x \in K(\psi, c_2, c_3), \psi(x) > c_2\}$. Then, for $x \in K(\psi, c_2, c_3)$, from (H_{10}), we have

$$\psi(A_\lambda x) = \min_{0 \le t \le T} \lambda \int_t^{t+T} G(t,s) f(s, x(h(s)))\, ds$$

$$\ge \frac{\lambda\delta}{1-\delta} \int_0^T \frac{c_2}{\delta}\, ds$$

$$\ge \frac{\delta}{1-\delta} \frac{c_2 T}{\delta} \lambda$$

$$> c_2.$$

Next, for $x \in \overline{K}_{c_1}$, we have

$$\|A_\lambda x\| = \sup_{0 \le t \le T} \lambda \int_t^{t+T} G(t,s) f(s, x(h(s))) \, ds$$

$$\le \frac{\lambda}{1-\delta} \int_0^T (1-\delta)c_1 \, ds$$

$$\le \frac{\lambda}{1-\delta}(1-\delta)\, c_1 T$$

$$< c_1$$

from (H_{11}). Also, for $x \in K(\psi, c_2, c_4)$ and $\|A_\lambda x\| > c_3$, we have

$$c_3 < \|A_\lambda x\| \le \frac{\lambda}{1-\delta} \int_0^T f(s, x(h(s))) \, ds,$$

which gives

$$\psi(A_\lambda x) \ge \frac{\lambda \delta}{1-\delta} \int_0^T f(s, x(h(s))) \, ds$$

$$> \delta c_3$$

$$= c_2.$$

Hence, all the conditions of Theorem 1.2.2 are satisfied, so Eq. (2.3) has at least three positive T-periodic solutions. This proves the theorem. $\qquad\square$

The proofs of Theorems 2.2.2–2.2.4 below are similar to that of Theorem 2.2.1 and hence are omitted.

Theorem 2.2.2 *Let* $\tilde{f}^\infty < T$ *and assume there are constants* $0 < c_1 < c_2$ *such that*

(H_{12}) $f(t,x) \ge \frac{2(1-\delta)}{\delta} c_2$ *for* $x \in K$, $c_2 \le x \le \frac{c_2}{\delta}$ *and* $0 \le t \le T$

and

(H_{13}) $f(t,x) < \delta(1-\delta)c_1$ *for* $x \in K$, $0 \le x \le c_1$ *and* $0 \le t \le T$.

Then (2.3) *has at least three positive* T-*periodic solutions for*

$$\frac{1}{2T} < \lambda < \frac{1}{T}.$$

Theorem 2.2.3 *Let $\tilde{f}^{\infty} < T$, $\tilde{f}^{0} < T$, and assume that there exists a constant $c_2 > 0$ such that (H_{10}) holds. Then there exist at least three positive T-periodic solutions of Eq. (2.3) for*

$$\frac{1 - \delta}{T} < \lambda < \frac{1}{T}.$$

Theorem 2.2.4 *Let $\tilde{f}^{\infty} < T$, $\tilde{f}^{0} < T$, and (H_{12}) hold. Then there exist at least three positive T-periodic solutions of (2.3) for*

$$\frac{1}{2T} < \lambda < \frac{1}{T}.$$

Remark 2.2.1 We need to choose

$$c_4 > \max\left\{\frac{M\,T}{(1 - \delta)(T - \delta_1)}, \frac{c_2}{\delta}\right\}$$

in the proofs of Theorems 2.2.2–2.2.4 in order to show that $A_\lambda : \overline{K}_{c_4} \to \overline{K}_{c_4}$, where M is given in the proof of Theorem 2.2.1.

Remark 2.2.2 It follows from the range on λ in Theorems 2.2.1–2.2.4 that λ and T depend on each other, that is, $\lambda T < 1$. Hence, if we consider the particular case $\lambda = 1$, then $T < 1$. This observation leads to the following theorem.

Theorem 2.2.5 *Let $\lambda \equiv 1$, $\tilde{f}^{\infty} < 1$, and $\tilde{f}^{0} < 1$. Assume that there exists a constant $c_2 > 0$ such that*

(H_{14}) $f(t, x) \geq a(t)c_2$ *for $x \in K$, $c_2 \leq x \leq \frac{c_2}{\delta}$ and $0 \leq t \leq T$.*

Then (2.3) has at least three positive T-periodic solutions.

Proof From the fact that $\tilde{f}^{\infty} < 1$, we can find $\sigma_1 \in (0, 1)$ and $\xi > 0$ such that $f(t, x) < \sigma_1 a(t)x$ for $x \geq \xi$. Let $\max\limits_{0 \leq x \leq \xi, 0 \leq t \leq T} f(t, x) = \eta$. Then $f(t, x) < \sigma_1 a(t)x + \eta$ for $x \geq 0$ and $0 \leq t \leq T$. Set

$$c_4 > \max\left\{\frac{\eta T}{(1 - \delta)(1 - \sigma_1)}, \frac{c_2}{\delta}\right\}.$$

Then, using the property that $\int_t^{t+T} G(t, s)a(s)\,ds \equiv 1$, it can easily be shown that $A_\lambda : \overline{K}_{c_4} \to \overline{K}_{c_4}$. Defining a nonnegative concave continuous functional by $\psi(x) = \min\limits_{t \in [0,T]} x(t)$ and a constant $c_3 = \frac{c_2}{\delta}$, (H_{14}) yields

$$\psi(A_\lambda x) = \min_{0 \le t \le T} \int_t^{t+T} G(t, s) f(s, x(h(s))) \, ds$$

$$\ge \min_{0 \le t \le T} \int_t^{t+T} G(t, s) a(s) c_2 \, ds$$

$$\ge c_2.$$

Next, since $\tilde{f}^0 < 1$, there exists a $\xi_1, 0 < \xi_1 < c_2$ such that

$$f(t, x) < a(t)x \quad \text{for } 0 < x < \xi_1.$$

Set $\xi_1 = c_1 < c_2$; then for $x \in \overline{K}_{c_1}$, we have

$$\|A_\lambda x\| = \sup_{0 \le t \le T} \int_t^{t+T} G(t, s) f(s, x(h(s))) \, ds$$

$$< \sup_{0 \le t \le T} \int_t^{t+T} G(t, s) a(s) x \, ds$$

$$< c_1.$$

Property (iii) of Theorem 1.2.2 is easy to verify. This shows that (2.3) has at least three positive T-periodic solutions and proves the theorem.   □

The next corollary follows from Theorem 2.2.5.

Corollary 2.2.1 *Let $\lambda \equiv 1$, $\tilde{f}^0 = 0$, $\tilde{f}^\infty = 0$, and there exists a constant $c_2 > 0$ such that (H_{14}) holds. Then (2.3) has at least three positive T-periodic solutions.*

Remark 2.2.3 Results derived in this section can be extended to Eq. (2.9) with $\lambda = 1$, that is, the equation

$$x'(t) = a(t)x(t) - f(t, x(t - \tau_1(t)), \dots, x(t - \tau_m(t))), \tag{2.23}$$

where $a, \tau_i, 1 \le i \le m$, and f are as defined in (2.9). Zhang et al. [10] obtained several sufficient conditions for the existence of at least two positive periodic solutions of (2.23) using Krasnosel'skii's fixed point theorem [1, 4]. On the other hand, the above results explain the existence of at least three positive periodic solutions under the same sufficient conditions.

2.3 Positive Periodic Solutions of the Equation $x'(t) = a(t)x(t) - \lambda b(t) f(t, x(h(t)))$

In this section, sufficient conditions are obtained for the existence of at least three positive T-periodic solutions of Eq. (2.4). Observe that (2.4) is equivalent to the integral equation

$$x(t) = \lambda \int_{t}^{t+T} G(t, s)b(s)f(s, x(h(s)))\, ds,$$

where $G(t, s)$ given in (2.20) satisfies (2.21). Consider the Banach space X as in (2.14) and a cone K as in (2.22) and define an operator A_λ on X by

$$(A_\lambda x)(t) = \lambda \int_{t}^{t+T} G(t, s)b(s)f(s, x(h(s)))\, ds.$$

It is easy to show that $A_\lambda : K \to K$ is completely continuous and the existence of a positive periodic solution of Eq. (2.4) is equivalent to the existence of a fixed point of the operator A_λ in K.

As above, the Leggett-Williams multiple fixed point Theorem 1.2.2 can be used to prove our results. Moreover, the results hold true if $b(t) \equiv 1$. In this case, the ranges on λ in the results obtained in this section are different from the ones given in Sect. 2.2.

Theorem 2.3.1 *Let $f^\infty < 1$ hold and assume that there are constants $0 < c_1 < c_2$ such that*

(H_{15}) $f(t, x) \geq \frac{2c_2}{\delta}$ *for* $x \in K$, $c_2 \leq x \leq \frac{c_2}{\delta}$ *and* $0 \leq t \leq T$

and

(H_{16}) $f(t, x) < c_1$ *for* $x \in K, 0 \leq x \leq c_1$ *and* $0 \leq t \leq T$.

Then (2.4) has at least three positive T-periodic solutions for

$$\frac{1 - \delta}{T \atop 2 \int\limits_{0}^{} b(t)\, dt} < \lambda < \frac{1 - \delta}{T \atop \int\limits_{0}^{} b(t)\, dt}.$$

Proof Now $f^\infty < 1$, so there exist $\sigma \in (0, 1)$ and $\xi > 0$ such that $f(t, x) \leq \sigma x$ for $x \geq \xi$ and $0 \leq t \leq T$. If

$$M = \max_{0 \leq x \leq \xi, 0 \leq t \leq T} f(t, x),$$

then $f(t, x) \leq \sigma x + M$ for $x \geq 0$ and $0 \leq t \leq T$. Choose $c_4 > 0$ such that

$$c_4 > \max \left\{ \frac{M}{1 - \sigma}, \frac{c_2}{\delta} \right\} > 0.$$

If $x \in \overline{K}_{c_4}$, we have

$$\|A_\lambda x\| = \sup_{0 \leq t \leq T} \lambda \int_t^{t+T} G(t, s) b(s) f(s, x(h(s))) \, ds$$

$$\leq \frac{\lambda}{1 - \delta} \int_0^T b(s) f(s, x(h(s))) \, ds$$

$$\leq \frac{\lambda}{1 - \delta} \int_0^T b(s)(\sigma x(h(s)) + M) \, ds$$

$$\leq \frac{\lambda}{1 - \delta} \int_0^T b(s)(\sigma \|x\| + M) \, ds$$

$$\leq \frac{\lambda(\sigma c_4 + M)}{1 - \delta} \int_0^T b(s) \, ds$$

$$\leq (\sigma c_4 + M)$$

$$< c_4.$$

This shows that $A_\lambda : \overline{K}_{c_4} \to \overline{K}_{c_4}$.

Now, define a nonnegative continuous concave functional ψ on K by $\psi(x) = \min_{t \in [0,T]} x(t)$. Let $c_3 = \frac{c_2}{\delta}$ and $\phi_0(t) = \phi_0$ be any given number satisfying $c_2 < \phi_0 < c_3$. Then $\phi_0 \in \{x; x \in K(\psi, c_2, c_3), \psi(x) > c_2\}$. For $x \in K(\psi, c_2, c_3)$, (H_{15}) gives

$$\psi(A_\lambda x) \geq \frac{\lambda \delta}{1 - \delta} \int_0^T b(s) f(s, x(h(s))) \, ds$$

$$\geq \frac{\lambda \delta}{1 - \delta} \frac{2 c_2}{\delta} \int_0^T b(s) \, ds$$

$$> c_2.$$

Next, for $x \in \overline{K}_{c_1}$, (H_{16}) implies

$$\|A_\lambda x\| \leq \frac{\lambda}{1-\delta} \int_0^T c_1 b(s)\, ds$$

$$< c_1.$$

For $x \in K(\psi, c_2, c_4)$ with $\|A_\lambda x\| > c_3$, we have

$$c_3 < \|A_\lambda x\| \leq \frac{\lambda}{1-\delta} \int_0^T b(s) f(s, x(h(s)))\, ds,$$

which, in turn implies that

$$\psi(A_\lambda x) \geq \frac{\lambda \delta}{1-\delta} \int_0^T b(s) f(s, x(h(s)))\, ds$$

$$> \delta c_3$$

$$= c_2.$$

Hence, by Theorem 1.2.2, Eq. (2.4) has at least three positive T-periodic solutions. This completes the proof of the theorem. □

Theorem 2.3.2 *Let* $f^\infty < 1$ *and* $f^0 < 1$. *Assume that there exists a constant* $c_2 > 0$ *such that* (H_{15}) *holds. Then Eq. (2.4) has at least three positive T-periodic solutions for*

$$\frac{1-\delta}{2\int_0^T b(t)\, dt} < \lambda < \frac{1-\delta}{\int_0^T b(t)\, dt}.$$

Proof We may proceed along the lines of the proof of Theorem 2.3.1 to prove this result. However, the following argument is needed to show that condition (ii) of Theorem 1.2.2 holds.

Since $f^0 < 1$, there exists a $\xi_1 \in (0, \frac{c_2}{2})$ such that $f(t, x) < x$ for $0 \leq x \leq \xi_1$. Choosing $c_1 = \xi_1$, we observe that $f(t, x) < x$ for $0 \leq x \leq c_1$ and $c_1 < c_2$. Then, for $x \in \overline{K}_{c_1}$,

$$\|A_\lambda x\| \leq \frac{\lambda}{1-\delta} \int_0^T b(s) x(h(s))\, ds$$

$$\leq \frac{\lambda}{1-\delta} \int_0^T b(s) \|x\|\, ds$$

$$\leq \frac{\lambda}{1-\delta} c_1 \int\limits_0^T b(s)\, ds$$

$$< c_1. \qquad\qquad\qquad \square$$

Theorem 2.3.3 *Let $f^\infty < T$ and assume there are constants $0 < c_1 < c_2$ such that*

(H_{17}) $f(t, x) \geq \frac{2T c_2}{\delta}$ *for $x \in K$, $c_2 \leq x \leq \frac{c_2}{\delta}$ and $0 \leq t \leq T$*

and

(H_{18}) $f(t, x) < c_1 T$ *for $x \in K$, $0 \leq x \leq c_1$ and $0 \leq t \leq T$.*

Then (2.4) has at least three positive T-periodic solutions for

$$\frac{1-\delta}{2T \int\limits_0^T b(t)\, dt} < \lambda < \frac{1-\delta}{T \int\limits_0^T b(t)\, dt}.$$

Proof Since $f^\infty < T$, there exist $\epsilon \in (0, T)$ and $\xi > 0$ such that $f(t, x) < \epsilon x$ for $x \geq \xi$. Let

$$M = \max_{0 \leq x \leq \xi, 0 \leq t \leq T} f(t, x).$$

Then $f(t, x) < \epsilon x + M$ for $x \geq 0$. Choose $c_4 > 0$ such that

$$c_4 > \max\left\{ \frac{M}{T-\epsilon}, \frac{c_2}{\delta} \right\}.$$

Now for $x \in \overline{K}_{c_4}$, we have

$$\|A_\lambda x\| = \sup_{0 \leq t \leq T} \lambda \int\limits_t^{t+T} G(t, s) b(s) f(s, x(h(s)))\, ds$$

$$\leq \frac{\lambda}{1-\delta} \int\limits_0^T b(s) f(s, x(h(s)))\, ds$$

$$\leq \frac{\lambda}{1-\delta} \int\limits_0^T b(s)(\epsilon x(h(s)) + M)\, ds$$

$$\leq \frac{\lambda}{1-\delta} \int\limits_0^T b(s)(\epsilon \|x\| + M)\, ds$$

$$\leq \frac{\lambda(\epsilon c_4 + M)}{1 - \delta} \int_0^T b(s)\, ds$$

$$\leq \frac{(\epsilon c_4 + M)}{T}$$

$$< c_4.$$

This shows that $A_\lambda : \overline{K}_{c_4} \to \overline{K}_{c_4}$.

Define ψ on K by $\psi(x) = \min_{t \in [0,T]} x(t)$ and let $c_3 = \frac{c_2}{\delta}$ and $\phi_0(t) = \phi_0$ be any given number satisfying $c_2 < \phi_0 < c_3$. Then $\phi_0 \in \{x \in K(\psi, c_2, c_3) : \psi(x) > c_2\}$. From (H_{17}), for $x \in K(\psi, c_2, c_3)$, we have

$$\psi(A_\lambda x) \geq \frac{\lambda \delta}{1 - \delta} \int_0^T b(s) f(s, x(h(s)))\, ds$$

$$\geq \frac{\lambda \delta}{1 - \delta} \frac{2 T c_2}{\delta} \int_0^T b(s)\, ds$$

$$> c_2.$$

Next for $x \in \overline{K}_{c_1}$, using (H_{18}) we obtain

$$\|A_\lambda x\| < \frac{\lambda}{1 - \delta} c_1 T \int_0^T b(s)\, ds$$

$$< c_1.$$

For $x \in K(\psi, c_2, c_4)$ with $\|A_\lambda x\| > c_3$, we have

$$c_3 < \|A_\lambda x\| \leq \frac{\lambda}{1 - \delta} \int_0^T b(s) f(s, x(h(s)))\, ds,$$

which, in turn implies that

$$\psi(A_\lambda x) \geq \frac{\lambda \delta}{1 - \delta} \int_0^T b(s) f(s, x(h(s)))\, ds$$

$$> \delta c_3$$

$$= c_2.$$

Hence, by Theorem 1.2.2, Eq. (2.4) has at least three positive T-periodic solutions and this completes the proof of the theorem. □

Theorem 2.3.4 Let $f^\infty < T$, and $f^0 < T$, and assume that there exists a constant $c_2 > 0$ such that (H_{17}) holds. Then (2.4) has at least three positive T-periodic solutions for

$$\frac{1-\delta}{2T \int_0^T b(t)\,dt} < \lambda < \frac{1-\delta}{T \int_0^T b(t)\,dt}.$$

Corollary 2.3.1 Let $f^\infty = 0$ and $f^0 = 0$. Assume that there exists a constant $c_2 > 0$ such that (H_{15}) holds. Then (2.4) has at least three positive T-periodic solutions for

$$\frac{1-\delta}{2 \int_0^T b(t)\,dt} < \lambda < \frac{1-\delta}{\int_0^T b(t)\,dt}.$$

Corollary 2.3.2 Let $f^\infty = 0$ and $f^0 = 0$. Suppose that there exists a constant $c_2 > 0$ such that (H_{17}) holds. Then there exist at least three positive T-periodic solutions of (2.4) for

$$\frac{1-\delta}{2T \int_0^T b(t)\,dt} < \lambda < \frac{1-\delta}{T \int_0^T b(t)\,dt}.$$

Remark 2.3.1 (Han and Wang [3]) obtained the following sufficient condition for the existence of two positive periodic solutions for the state-dependent delay differential equation

$$x'(t) = a(t, x(t))x(t) - f(t, x(t - \tau_1(t, x(t))), \ldots, x(t - \tau_m(t, x(t)))) \quad (2.24)$$

using fixed point theorem in cones [1]. They assumed that $a \in C(R \times R_+, R)$, $a(t + T, x) = a(t, x)$ for any $(t, x) \in R \times R_+$, $f \in C(R \times [R]^m, R_+)$, $f(t + T, x_1, \ldots, x_m) = f(t, x_1, \ldots, x_m)$, $\tau_i(t + T, x) = \tau_i(t, x)$ for any $x \in R_+, t \in R$, $i = 1, \ldots, m$, and $T > 0$ is a constant.

Theorem 2.3.5 Han and Wang [3] *Assume that* $a_1(t) \leq a(t, x) \leq a_2(t)$ *for any* $(t, x) \in R \times R_+$, *where* a_1 *and* a_2 *are nonnegative* T-*periodic continuous functions on* R *and* $\int_0^T a_1(s)\,ds > 0$. *Let*

$$\limsup_{|u| \to +0} \frac{f(t, u_1, u_2, \ldots, u_m)}{|u|} < \frac{a_2(t)}{\gamma}$$

and

$$\limsup_{|u| \to +\infty} \frac{f(t, u_1, u_2, \ldots, u_m)}{|u|} < \frac{a_2(t)}{\gamma}$$

uniformly for $t \in R$, where $\gamma = \dfrac{\exp(\int_0^T a_2(t)dt)-1}{\exp(\int_0^T a_1(t)dt)-1}$. Next, suppose that there exists a
$\rho > 0$ such that the inequality $\sigma\rho \le |u| \le \rho$ yields $f(t, u_1, u_2, \ldots, u_m) > a_1(t)\rho\gamma$
for $t \in [0, T]$, where $|u| = \max_i\{u_1, u_2, \ldots, u_m\}$ and

$$\sigma = \frac{\left(\inf_{0 \le t \le s \le T} \exp\left(\int_t^s a_1(\theta)\,d\theta\right)\right)\left(\exp\left(\int_0^T a_1(\theta)\,d\theta\right) - 1\right)}{\left(\sup_{0 \le t \le s \le T} \exp\left(\int_t^s a_2(\theta)\,d\theta\right)\right)\left(\exp\left(\int_0^T a_2(\theta)\,d\theta\right) - 1\right)}.$$

Then (2.24) has at least two positive periodic solutions x_1 and x_2 such that $0 < \|x_1\| < \rho < \|x_2\|$.

With $\lambda \equiv 1$, Eq. (2.4) becomes

$$x'(t) = a(t)x(t) - F(t, x(h(t))), \qquad (2.25)$$

where $F(t, x) = b(t)f(t, x)$. Applying Theorem 2.3.5 to Eq. (2.25) gives the following result.

Theorem 2.3.6 *Let $\limsup_{x \to 0} \dfrac{F(t,x)}{x} < a(t)$ and $\limsup_{x \to \infty} \dfrac{F(t,x)}{x} < a(t)$ for $0 \le t \le T$ and assume there exists $\rho > 0$ such that*

(H_{19}) *$F(t, x) > \rho a(t)$ for $0 \le t \le T$ and $\delta\rho \le x \le \rho$,*

where $\delta = \exp(-\int_0^T a(\theta)\,d\theta)$. Then (2.25) has at least two positive T-periodic solutions x_1 and x_2 such that $0 < \|x_1\| < \rho < \|x_2\|$.

Corollary 2.3.3 *Let $\limsup_{x \to 0} \dfrac{F(t,x)}{x} = 0$, $\limsup_{x \to \infty} \dfrac{F(t,x)}{x} = 0$, and assume there exists $\rho > 0$ such that (H_{19}) holds. Then (2.25) has at least two positive periodic solutions.*

If we ask that $\frac{1}{2} < \dfrac{\int_0^T b(t)\,dt}{1-\delta} < 1$, then from Corollary 2.3.1 we have the following result.

Corollary 2.3.4 *Let $f^\infty = 0$ and $f^0 = 0$. Assume there exists a constant $c_2 > 0$ such that (H_{15}) holds. Then Eq. (2.25) has at least three positive T-periodic solutions.*

Similarly, if we assume $\frac{1}{2} < \dfrac{T\int_0^T b(t)\,dt}{1-\delta} < 1$, then the following corollary follows from Corollary 2.3.2.

Corollary 2.3.5 *Let $f^\infty = 0$ and $f^0 = 0$. Assume there exists a constant $c_2 > 0$ such that (H_{17}) holds. Then Eq. (2.25) has at least three positive T-periodic solutions.*

Corollaries 2.3.1–2.3.2 and 2.3.4–2.3.5 extend and improve Corollary 2.3.3. In fact, under very similar condition, Corollary 2.3.3 and Corollary 2.3.4 yield that (2.25) has at least three positive periodic solutions.

Example 2.3.1 Consider

$$x'(t) = \frac{1}{4\pi}\left(\frac{3}{2} + \sin^2 t\right)x(t) - \frac{1}{20\pi}(1 + \cos^2 t)e^5 x^2(t - \tau)e^{-x(t-\tau)}, \quad t \geq 0,$$
(2.26)

where $\tau > 0$ is a constant. Here $a(t) = \frac{1}{4\pi}(\frac{3}{2} + \sin^2 t)$, $b(t) = 1 + \cos^2 t$, $T = \pi$, $\delta = e^{-\int_0^\pi a(s)\,ds} = e^{-1/2}$ and $\int_0^\pi b(t)\,dt = \frac{3\pi}{2}$. Set $f(t, x) = \frac{1}{\pi}e^5 x^2 e^{-x}$ and $\lambda = \frac{1}{20} = 0.05$. Then $f^\infty = 0 < 1$ and $0.04 = \frac{1-\delta}{2\int_0^T b(t)\,dt} < 0.05 = \lambda < 0.08 = \frac{1-\delta}{\int_0^T b(t)\,dt}$. Set $c_2 = 2$ and $c_1 = 0.02$. Clearly, $f(t, x) = \frac{1}{\pi}e^5 x^2 e^{-x} > \frac{1}{\pi}e^5 c_2^2 e^{-c_2 e^{1/2}}$. Now, for $c_2 = 2$, we observe that $\frac{1}{\pi}e^5 c_2^2 e^{-c_2 e^{1/2}} > 2c_2 e^{1/2}$. This in turn implies that (H_{15}) holds. Since $f(t, x) = \frac{1}{\pi}e^5 x^2 e^{-x} < \frac{1}{\pi}e^5 c_1^2$ for $0 \leq x \leq c_1$, condition (H_{16}) is satisfied for $c_1 = 0.02$. Also, $0 < c_1 < c_2$. Thus, by Theorem 2.3.1, Eq. (2.26) has at least three positive π-periodic solutions.

Example 2.3.2 By Theorem 2.3.3, the equation

$$x'(t) = \frac{1}{2\pi}(1 + \cos^2 t)x(t) - \frac{1}{50\pi}(1 + \sin^2 t)e^6 x^2(t - \tau)e^{-x(t-\tau)}, \quad t \geq 0$$

has at least three positive π-periodic solutions, where $\tau > 0$ is a constant. Here, we need to choose $c_2 = 1$ and $c_1 = 0.024$.

2.4 Periodic Solutions of State-Dependent Differential Equations

Consider the state-dependent delay differential equation

$$x'(t) = -a(t, x(t))x(t) + \lambda f(t, x(t - \tau_1(t, x(t))), \ldots, x(t - \tau_m(t, x(t))))$$
(2.27)

where $\lambda > 0$ is a parameter, $T > 0$ is a constant, $a \in C(R \times R_+, R)$, $a(t + T, x) = a(t, x)$ for any $(t, x) \in R \times R_+$, $f \in C(R \times R_+, R)$, $f(t + T, x_1, x_2, \ldots, x_m) = f(t, x_1, x_2, \ldots, x_m)$, and $\tau_i(t + T, x) = \tau_i(t, x)$ for any $x \in R_+$, $t \in R$, and $i = 1, 2, \ldots, m$. We assume that there exist two nonnegative T-periodic functions $b(t)$ and $c(t)$ such that $b(t) \leq a(t, x) \leq c(t)$ for any $(t, x) \in R \times R_+$ and $\int_0^T b(t)\,dt > 0$.

We wish to point out that the method applied in this section can also be used to obtain similar results for state-dependent delay differential equations of the form (2.24).

If $a(t, x) = a(t)g(x(t))$ and $\tau_i(t, x(t)) = \tau_i(t)$, $i = 1, 2, \ldots, n$, where $g \in C([0, \infty), [0, \infty))$, then Eqs. (2.27) and (2.24) take the forms

$$x'(t) = -a(t)g(x(t))x(t) + \lambda f(t, x(t - \tau_1(t)), x(t - \tau_2(t)), \ldots, x(t - \tau_m(t)))$$
(2.28)

and

$$x'(t) = a(t)g(x(t))x(t) - \lambda f(t, x(t-\tau_1(t)), x(t-\tau_2(t)), \ldots, x(t-\tau_m(t))), \quad (2.29)$$

respectively.

Let $X = \{x(t) : x(t + T) = x(t), t \in R\}$ and $\|x\| = \max_{0 \le t \le T} |x(t)|$. Then X is a Banach space endowed with the norm $\|\cdot\|$. Clearly, x is a positive T-periodic solution of (2.27) if and only if x is a T-periodic solution of the integral equation

$$x(t) = \lambda \int_t^{t+T} G(t, s) f(s, x(s - \tau_1(s, x(s))), \ldots, x(s - \tau_m(s, x(s)))) \, ds,$$

where

$$G(t, s) = \frac{\exp\left(\int_t^s a(\theta, x(\theta)) \, d\theta\right)}{\exp\left(\int_0^T a(\theta, x(\theta)) \, d\theta\right) - 1}.$$

In view of the above, we define an operator A_λ by

$$A_\lambda x = \lambda \int_t^{t+T} G(t, s) f(s, x(s - \tau_1(s, x(s))), \ldots, x(s - \tau_m(s, x(s)))) \, ds \quad (2.30)$$

for every $x \in X$ and $t \in R$. Clearly $A_\lambda x(t + T) = A_\lambda x(t)$ and $A_\lambda : X \to X$. The Green's kernel $G(t, s)$ satisfies the inequality

$$\alpha = \frac{1}{\exp\left(\int_0^T c(\theta) \, d\theta\right) - 1} \le |G(t, s)| \le \frac{\exp\left(\int_0^T c(\theta) \, d\theta\right)}{\exp\left(\int_0^T b(\theta) \, d\theta\right) - 1} = \beta$$

for every $0 \le t \le s \le t + T$. Let $k_1 = \exp\left(\int_0^T b(\theta) \, d\theta\right)$ and $k_2 = \exp\left(\int_0^T c(\theta) \, d\theta\right)$. Then,

$$\alpha = \frac{1}{k_2 - 1}, \quad \beta = \frac{k_2}{k_1 - 1}, \quad k_1 \le k_2, \quad \text{and} \quad \delta = \frac{\beta}{\alpha} = \frac{k_2(k_2 - 1)}{(k_1 - 1)} > 1.$$

For any $x \in X$, we have

$$\|A_\lambda x\| \leq \frac{\lambda k_2}{k_1 - 1} \int_0^T f(s, x(s - \tau_1(s, x(s))), \ldots, x(s - \tau_m(s, x(s)))) \, ds$$

and

$$(A_\lambda x)(t) \geq \frac{\lambda}{k_2 - 1} \int_0^T f(s, x(s - \tau_1(s, x(s))), \ldots, x(s - \tau_m(s, x(s)))) \, ds$$

$$\geq \frac{(k_1 - 1)}{k_2(k_2 - 1)} \|A_\lambda x\| = \frac{1}{\delta} \|A_\lambda x\|.$$

Thus, if we define a cone K on X by

$$K = \left\{ x \in X : x(t) \geq \frac{1}{\delta} \|x\| \right\},$$

then $A_\lambda : K \to K$. It is easy to show that $A_\lambda : K \to K$ is completely continuous.
Define

$$f^\theta = \limsup_{|x| \to \theta} \max_{0 \leq t \leq T} \frac{f(t, x)}{c(t)|x|},$$

where $|x| = \max_{1 \leq i \leq m}\{x_1, x_2, \ldots, x_m\}$.

Theorem 2.4.1 *Let $f^0 < T$, $f^\infty < T$, and assume that there exists a constant $c_2 > 0$ such that*

$$f(t, x_1, x_2, \ldots, x_m) \geq 2T k_1 \left(\frac{k_2 - 1}{k_1 - 1}\right)^2 b(t)c_2 \quad \text{for } x \in K \text{ and } c_2 \leq |x| \leq \delta c_2. \tag{2.31}$$

Then Eq. (2.27) has at least three positive T-periodic solutions for

$$\frac{1}{2T} \frac{k_1 - 1}{k_2 - 1} \leq \lambda \leq \frac{1}{T} \frac{k_1 - 1}{k_2 - 1}.$$

Proof First, suppose that $f^\infty < T$. Then there exist $0 < \epsilon < T$ and $c_3 = \delta c_2 > c_2$ such that

$$f(t, x_1, x_2, \ldots, x_m) < c(t)(T - \epsilon)|x| \quad \text{for } |x| > c_3 \text{ and } t \in R.$$

Set $c_4 = \delta c_3$. Clearly, $c_4 \geq \|x\| \geq x \geq \frac{1}{\delta}\|x\|$ for $x \in K \cap \overline{K}_{c_4}$. For $x \in \overline{K}_{c_4}$,

$$\|A_\lambda x\| = \lambda \int_t^{t+T} G(t, s) f(s, x(s - \tau_1(s, x(s))), \ldots, x(s - \tau_m(s, x(s)))) \, ds$$

$$\leq \lambda(T - \epsilon) \int_t^{t+T} G(t,s)c(s) \max_{1 \leq i \leq m} |x(s - \tau_i(s, x(s)))| \, ds$$

$$\leq \lambda(T - \epsilon)c_4 \int_t^{t+T} \frac{\exp\left(\int_t^s c(\theta) \, d\theta\right)}{\exp\left(\int_0^T b(\theta) \, d\theta\right) - 1} c(s) \, ds$$

$$\leq \lambda T c_4 \frac{k_2 - 1}{k_1 - 1} \leq c_4.$$

Since $A_\lambda : K \to K$, then, in addition to the above, it follows that $A_\lambda : \overline{K}_{c_4} \to \overline{K}_{c_4}$. Define a nonnegative continuous function ψ on K by $\psi(x) = \min_{t \in [0,T]} |x(t)|$. Then $\psi(x) \leq \|x\|$. Let $\phi_0(t) = \phi_0$, where ϕ_0 is any given number satisfying $c_2 < \phi_0 < c_3$. Then, $\phi_0 \in \{x \in K(\psi, c_2, c_3) : \psi(x) > c_2\} \neq \phi$. For $x \in K(\psi, c_2, c_3)$, using (2.31) we obtain

$$\psi(A_\lambda x) = \min_{0 \leq t \leq T} \lambda \int_t^{t+T} G(t,s) f(s, x(s - \tau_1(s, x(s))), \ldots, x(s - \tau_m(s, x(s)))) \, ds$$

$$\geq 2Tk_1 \left(\frac{k_2 - 1}{k_1 - 1}\right)^2 c_2 \lambda \int_t^{t+T} \frac{1}{\exp\left(\int_0^T c(\theta) \, d\theta\right) - 1} b(s) \, ds$$

$$\geq 2Tk_1 \left(\frac{k_2 - 1}{k_1 - 1}\right)^2 c_2 \lambda \int_t^{t+T} \frac{\exp\left(\int_t^s b(\theta) \, d\theta\right) b(s)}{\exp\left(\int_t^s b(\theta) \, d\theta\right)\left(\exp\left(\int_0^T c(\theta) \, d\theta\right) - 1\right)} \, ds$$

$$\geq 2Tk_1 \left(\frac{k_2 - 1}{k_1 - 1}\right)^2 c_2 \lambda \frac{1}{\exp\left(\int_0^T b(\theta) \, d\theta\right)} \int_t^{t+T} \frac{\exp\left(\int_t^s b(\theta) \, d\theta\right) b(s)}{\exp\left(\int_0^T c(\theta) \, d\theta\right) - 1} \, ds$$

$$\geq 2Tk_1 \left(\frac{k_2 - 1}{k_1 - 1}\right)^2 c_2 \lambda \frac{1}{k_1} \cdot \frac{k_1 - 1}{k_2 - 1} \geq c_2.$$

Hence, condition (i) of Theorem 1.2.2 is satisfied.

Now $f^0 < T$ implies there are $\epsilon_1 > 0$ and $c_1 < c_2$ such that

$$f(t, x_1, x_2, \ldots, x_m) < c_1(T - \epsilon_1)c(t)|x| \quad for \quad 0 < |x| < c_1.$$

Since $c_1 \geq \|x\| \geq x(t) \geq \frac{1}{\delta}\|x\|$ for any $x \in K \cap \overline{K}_{c_1}$, for any $x \in \overline{K}_{c_1}$, we have

$$\|A_\lambda x\| \leq \lambda \int_t^{t+T} G(t,s)(T-\epsilon_1)c(s) \max_{0 \leq i \leq m} |x(s-\tau_i(s,x(s)))| \, ds$$

$$\leq \lambda(T-\epsilon_1)c_1 \int_t^{t+T} \frac{\exp\left(\int_t^s c(\theta)\,d\theta\right) c(s)}{\exp\left(\int_0^T b(\theta)\,d\theta\right) - 1} \, ds$$

$$\leq \lambda T c_1 \left(\frac{k_2-1}{k_1-1}\right) \leq c_1.$$

Thus, condition (ii) of Theorem 1.2.2 holds.

Finally, for any $x \in K(\psi, c_2, c_4)$ and $\|A_\lambda x\| > c_3$, we see that

$$c_3 \leq \|A_\lambda x\| \leq \beta\lambda \int_t^{t+T} f(s, x(s-\tau_1(s,x(s))), \ldots, x(s-\tau_m(s,x(s)))) \, ds,$$

and it follows that

$$\psi(A_\lambda x) \geq \alpha\lambda \int_t^{t+T} G(t,s) f(s, x(s-\tau_1(s,x(s))), \ldots, x(s-\tau_m(s,x(s)))) \, ds$$

$$\geq \frac{\alpha}{\beta}c_3 = \frac{c_3}{\delta} = c_2.$$

Hence, by Theorem 1.2.2, Eq. (2.27) has at least three positive T-periodic solutions. This completes the proof of the theorem. □

The following theorem follows from the proof of Theorem 2.4.1.

Theorem 2.4.2 *Let $f^0 < 1$, $f^\infty < 1$, and assume there exists $c_2 > 0$ such that*

$$f(t, x_1, x_2, \ldots, x_m) \geq 2k_1 \left(\frac{k_2-1}{k_1-1}\right)^2 b(t)c_2 \text{ for } x \in K \text{ and } c_2 \leq |x| \leq \delta c_2.$$

Then Eq. (2.27) has at least three positive T-periodic solutions for

$$\frac{1}{2}\frac{k_1-1}{k_2-1} \leq \lambda \leq \frac{k_1-1}{k_2-1}.$$

Theorem 2.4.3 *Let $f^0 < T$, $f^\infty < T$, and assume there exists $c_2 > 0$ such that*

$$f(t, x_1, x_2, \ldots, x_m) \geq \frac{\beta}{\alpha} T k_1 \left(\frac{k_2-1}{k_1-1}\right)^2 b(t)c_2 \text{ for } x \in K \text{ and } c_2 \leq |x| \leq \delta c_2.$$

Then Eq. (2.27) has at least three positive T-periodic solutions for

$$\frac{\alpha}{\beta T}\frac{k_1 - 1}{k_2 - 1} \leq \lambda \leq \frac{1}{T}\frac{k_1 - 1}{k_2 - 1}.$$

Proof Choose c_3 and c_4 as in the proof of Theorem 2.4.1. Proceeding along the lines of that proof, we can show that $A_\lambda : \overline{K}_{c_4} \to \overline{K}_{c_4}$ and conditions (ii) and (iii) of Theorem 1.2.2 hold. In order to complete the proof of the theorem, we need to verify condition (i) of Theorem 1.2.2. Let $\phi_0(t) = \phi_0$, where ϕ_0 is any number satisfying $c_2 < \phi_0 < c_3$. Then, $\phi_0 \in \{x \in K(\psi, c_2, c_3) : \psi(x) > c_2\} \neq \phi$. For $x \in K(\psi, c_2, c_3)$, we have

$$\psi(A_\lambda x) = \min_{0 \leq t \leq T} \lambda \int_t^{t+T} G(t,s) f(s, x(s - \tau_1(s, x(s))), \ldots, x(s - \tau_m(s, x(s)))) \, ds$$

$$\geq \lambda T k_1 \frac{\beta}{\alpha} \left(\frac{k_2 - 1}{k_1 - 1}\right)^2 c_2 \int_t^{t+T} \frac{1}{\exp\left(\int_0^T c(\theta) \, d\theta\right) - 1} b(s) \, ds$$

$$\geq \lambda T k_1 \frac{\beta}{\alpha} \left(\frac{k_2 - 1}{k_1 - 1}\right)^2 c_2 \int_t^{t+T} \frac{\exp\left(\int_t^s b(\theta) \, d\theta\right) b(s)}{\exp\left(\int_t^s b(\theta) \, d\theta\right)\left(\exp\left(\int_0^T c(\theta) \, d\theta\right) - 1\right)} \, ds$$

$$\geq \lambda T k_1 \frac{\beta}{\alpha} \left(\frac{k_2 - 1}{k_1 - 1}\right)^2 c_2 \frac{1}{k_1} \int_t^{t+T} \frac{\exp\left(\int_t^s b(\theta) \, d\theta\right) b(s)}{\exp\left(\int_0^T c(\theta) \, d\theta\right) - 1} \, ds$$

$$\geq \lambda T \frac{\beta}{\alpha} \left(\frac{k_2 - 1}{k_1 - 1}\right)^2 c_2 \frac{k_1 - 1}{k_2 - 1} \geq c_2.$$

Hence, by Theorem 1.2.2, Eq. (2.27) has at least three positive T-periodic solutions, and this proves the theorem. □

The following theorem follows from the proof of Theorem 2.4.3.

Theorem 2.4.4 *Let $f^0 < 1$, $f^\infty < 1$, and assume there exists $c_2 > 0$ such that*

$$f(t, x_1, x_2, \ldots, x_m) \geq \frac{\beta}{\alpha} k_1 \left(\frac{k_2 - 1}{k_1 - 1}\right)^2 b(t) c_2 \quad \text{for } x \in K \text{ and } c_2 \leq |x| \leq \delta c_2.$$

Then Eq. (2.27) has at least three positive T-periodic solutions for

$$\frac{\alpha}{\beta}\frac{k_1 - 1}{k_2 - 1} \leq \lambda \leq \frac{k_1 - 1}{k_2 - 1}.$$

Theorem 2.4.5 *Let $f^0 < T$, $f^\infty < T$, and assume there exists $c_2 > 0$ such that*

$$f(t, x_1, x_2, \ldots, x_m) \geq T \frac{\beta^2}{\alpha^2} c_2 \int_0^T c(s)\, ds \quad \text{for } x \in K \text{ and } c_2 \leq |x| \leq \delta c_2.$$

Then Eq. (2.27) has at least three positive T-periodic solutions for

$$\frac{\alpha}{\beta^2 T \int_0^T c(s)\, ds} \leq \lambda \leq \frac{1}{\beta T \int_0^T c(s)\, ds}.$$

Proof Let c_3 and c_4 be as in the proof of Theorem 2.4.1. Following the proof of Theorem 2.4.1 with small modifications, it can be shown that $A_\lambda : \overline{K}_{c_4} \to \overline{K}_{c_4}$ and conditions (ii) and (iii) of Theorem 1.2.2 hold. Let $\phi_0(t) = \phi_0$, where ϕ_0 is any number satisfying $c_2 < \phi_0 < c_3$. Then $\phi_0 \in \{x \in K(\psi, c_2, c_3) : \psi(x) > c_2\} \neq \phi$. In order to apply Theorem 1.2.2, we only need to show that $\psi(A_\lambda x) > c_2$ for all $x \in K(\psi, c_2, c_3)$. Now for $x \in K(\psi, c_2, c_3)$,

$$\psi(A_\lambda x) \geq \lambda \alpha \int_t^{t+T} f(s, x(s - \tau_1(s, x(s))), \ldots, x(s - \tau_m(s, x(s))))\, ds$$

$$\geq \lambda \alpha \frac{\beta^2}{\alpha^2} T c_2 \int_0^T c(s)\, ds$$

$$\geq \frac{\alpha}{\beta^2 T \int_0^T c(s)\, ds} \cdot \alpha \frac{\beta^2}{\alpha^2} T c_2 \int_0^T c(s)\, ds = c_2.$$

By Theorem 1.2.2, Eq. (2.27) has at least three positive T-periodic solutions. The proof of the theorem is complete.   □

The proof of the following theorem should now be clear and we leave the details to the reader.

Theorem 2.4.6 *Let $f^0 < 1$, $f^\infty < 1$, and*

$$f(t, x_1, x_2, \ldots, x_m) \geq \frac{\beta^2}{\alpha^2} c_2 \int_0^T c(s)\, ds \quad \text{for } x \in K \text{ and } c_2 \leq |x| \leq \delta c_2.$$

Then Eq. (2.27) has at least three positive T-periodic solutions for

$$\frac{\alpha}{\beta^2 \int\limits_0^T c(s)\,ds} \le \lambda \le \frac{1}{\beta \int\limits_0^T c(s)\,ds}.$$

Example 2.4.1   Consider the differential equation

$$x'(t) = -\frac{\log 2}{2\pi}(2 + \sin t)\left(1 + \frac{1}{1+x(t)}\right)x(t) + \frac{1}{11\pi}\left(x(t) + 6e^{23}x^2(t)e^{-x(t)}\right).$$

$$(2.32)$$

Here $T = 2\pi$ and $a(t, x) = \frac{\log 2}{2\pi}(2 + \sin t)\left(1 + \frac{1}{1+x(t)}\right)$. Setting $b(t) = \frac{\log 2}{2\pi}(2 + \sin t)$ and $c(t) = \frac{\log 2}{\pi}(2 + \sin t)$, it follows that $0 < b(t) \le |a(t, x)| \le c(t)$ for $(t, x) \in R \times R_+$. Then $k_1 = 4$, $k_2 = 16$, $\left(\frac{k_2-1}{k_1-1}\right)^2 = 25$, and $\delta = 80$. Set $\lambda = \frac{1}{11\pi}$ and $f(t, x) = x + 6e^{23}x^2e^{-x}$. Then $f^0 < 2\pi$ and $f^\infty < 2\pi$. Clearly, $\frac{1}{2T}\frac{k_1-1}{k_2-1} \le \lambda \le \frac{1}{T}\frac{k_1-1}{k_2-1}$, that is, $0.0159 \le 0.02894 \le 0.03183$. To show that (2.32) has at least three positive T-periodic solutions, we choose $c_2 = \frac{1}{6}$. Then for $c_2 \le x \le \delta c_2$, that is, for $\frac{1}{6} \le x \le \frac{80}{6}$, we have

$$f(t, x) = x + 6e^{23}x^2e^{-x} \ge c_2 + 6e^{23}c_2^2e^{-\delta c_2}$$

$$\ge \frac{1}{6} + \frac{1}{6}e^{23-\frac{80}{6}}$$

$$\ge \frac{1}{6} \times 2 \times 2\pi \times 25 \times 4 \times \frac{3\log 2}{2\pi}$$

$$\ge 2Tk_1\left(\frac{k_2-1}{k_1-1}\right)^2 b(t)c_2.$$

Hence, by Theorem 2.4.1, Eq. (2.32) has at least three 2π-periodic solutions.

Now, we shall apply the previous theorems in this section to delay differential equations with a parameter of the form (2.28). Similar results can be obtained for (2.29).

We assume that $g \in C([0, \infty), [0, \infty))$ and there are constants $0 < l < L$ such that $l \le g(x) \le L$ for $x \ge 0$. Set $\sigma = \exp\left(\int_0^T a(\theta)\,d\theta\right)$, $k_1 = \sigma^l$ and $k_2 = \sigma^L$. We have the following relations:

$$\alpha = \frac{1}{\sigma^L - 1}, \quad \beta = \frac{\sigma^L}{\sigma^l - 1}, \quad \delta = \frac{\beta}{\alpha} = \sigma^L\frac{\sigma^L - 1}{\sigma^l - 1} \quad \text{and} \quad \frac{k_2 - 1}{k_1 - 1} = \frac{\sigma^L - 1}{\sigma^l - 1}.$$

Define

$$\tilde{f}^h = \limsup_{|x|\to h}\ \max_{0\le t\le T}\frac{f(t, x)}{a(t)|x|}.$$

Applying Theorems 2.4.1–2.4.6 to Eq. (2.28), we obtain the following results.

Theorem 2.4.7 *Let $\tilde{f}^0 < LT$, $\tilde{f}^\infty < LT$, and assume there is a constant $c_2 > 0$ such that*

$$f(t, x_1, x_2, \ldots, x_m) \geq 2Tl\sigma^l \left(\frac{\sigma^L - 1}{\sigma^l - 1}\right)^2 a(t)c_2 \quad \text{for } x \in K \text{ and } c_2 \leq |x| \leq \delta c_2.$$

Then Eq. (2.28) has at least three positive T-periodic solutions for

$$\frac{1}{2T} \frac{\sigma^l - 1}{\sigma^L - 1} \leq \lambda \leq \frac{1}{T} \frac{\sigma^l - 1}{\sigma^L - 1}.$$

Theorem 2.4.8 *Let $\tilde{f}^0 < L$, $\tilde{f}^\infty < L$, and assume there exists $c_2 > 0$ such that*

$$f(t, x_1, x_2, \ldots, x_m) \geq 2l\sigma^l \left(\frac{\sigma^L - 1}{\sigma^l - 1}\right)^2 a(t)c_2 \quad \text{for } x \in K \text{ and } c_2 \leq |x| \leq \delta c_2.$$

Then Eq. (2.28) has at least three positive T-periodic solutions for

$$\frac{1}{2} \frac{\sigma^l - 1}{\sigma^L - 1} \leq \lambda \leq \frac{\sigma^l - 1}{\sigma^L - 1}.$$

Theorem 2.4.9 *Let $\tilde{f}^0 < LT$, $\tilde{f}^\infty < LT$, and assume there exists $c_2 > 0$ such that*

$$f(t, x_1, x_2, \ldots, x_m) \geq \sigma^{L+l} T \left(\frac{\sigma^L - 1}{\sigma^l - 1}\right)^3 l\, a(t)c_2 \quad for \ x \in K \ and$$

$$c_2 \leq |x| \leq \delta c_2.$$

Then Eq. (2.28) has at least three positive T-periodic solutions for

$$\frac{1}{T\sigma^{L-l}} \frac{\sigma^l - 1}{\sigma^L - 1} \leq \lambda \leq \frac{1}{T} \frac{\sigma^l - 1}{\sigma^L - 1}.$$

Theorem 2.4.10 *Let $\tilde{f}^0 < L$, $\tilde{f}^\infty < L$, and assume there is a constant $c_2 > 0$ such that*

$$f(t, x_1, x_2, \ldots, x_m) \geq \sigma^{L+l} \left(\frac{\sigma^L - 1}{\sigma^l - 1}\right)^3 l\, a(t)c_2 \quad \text{for } x \in K \text{ and } c_2 \leq |x| \leq \delta c_2.$$

Then Eq. (2.28) has at least three positive T-periodic solutions for

$$\frac{1}{\sigma^{L-l}} \frac{\sigma^l - 1}{\sigma^L - 1} \leq \lambda \leq \frac{\sigma^l - 1}{\sigma^L - 1}.$$

Theorem 2.4.11 *Let $\tilde{f}^0 < LT$, $\tilde{f}^\infty < LT$, and assume there exists $c_2 > 0$ such that*

$$f(t, x_1, x_2, \ldots, x_m) \geq \sigma^{2L} c_2 T \left(\frac{\sigma^L - 1}{\sigma^l - 1}\right)^2 L \int\limits_0^T a(s)\, ds \quad \text{for } x \in K \text{ and}$$

$$c_2 \leq |x| \leq \delta c_2.$$

Then Eq. (2.28) has at least three positive T-periodic solutions for

$$\frac{(e^l - 1)^2}{T(e^L - 1)e^{2L}L \int\limits_0^T a(\theta)\, d\theta} \leq \lambda \leq \frac{(e^l - 1)}{Te^L L \int\limits_0^T a(\theta)\, d\theta}.$$

Theorem 2.4.12 *Let $\tilde{f}^0 < L$, $\tilde{f}^\infty < L$, and assume that there exists a positive constant $c_2 > 0$ such that*

$$f(t, x_1, x_2, \ldots, x_m) \geq \sigma^{2L} c_2 \left(\frac{\sigma^L - 1}{\sigma^l - 1}\right)^2 L \int\limits_0^T a(s)\, ds \quad \text{for } x \in K \text{ and}$$

$$c_2 \leq |x| \leq \delta c_2.$$

Then Eq. (2.28) has at least three positive T-periodic solutions for

$$\frac{(e^l - 1)^2}{(e^L - 1)e^{2L}L \int\limits_0^T a(\theta)\, d\theta} \leq \lambda \leq \frac{(e^l - 1)}{e^L L \int\limits_0^T a(\theta)\, d\theta}.$$

Now we direct our attention to the particular case when $a(t) \equiv a$ is a constant and $g(x) \equiv 1$ with $l = L = 1$. Then $\delta = \sigma = e^{aT}$ and $k_1 = k_2$. Let

$$f^{*\theta} = \frac{1}{a} \limsup_{|x| \to \theta} \max_{0 \leq t \leq T} \frac{f(t, x)}{|x|}.$$

Applying Theorems 2.4.7–2.4.12 to the equation

$$x'(t) = -ax(t) + \lambda f(t, x(t - \tau)), \tag{2.33}$$

we obtain the following interesting results.

Theorem 2.4.13 *Let $f^{*0} < aT$, $f^{*\infty} < aT$, and assume there exists a constant $c_2 > 0$ such that*

$$f(t, x) \geq 2a\delta c_2 T \quad \text{for } x \in K \text{ and } c_2 \leq |x| \leq \delta c_2.$$

Then Eq. (2.33) has at least three positive T-periodic solutions for

$$\frac{1}{2T} \leq \lambda \leq \frac{1}{T}.$$

Theorem 2.4.14 *Let* $f^{*0} < a$ *and* $f^{*\infty} < a$. *If there exists a constant* $c_2 > 0$ *such that*

$$f(t, x) \geq 2a\delta c_2 \quad \text{for } x \in K \text{ and } c_2 \leq |x| \leq \delta c_2,$$

then Eq. (2.33) has at least three positive T-periodic solutions for

$$\frac{1}{2} \leq \lambda \leq 1.$$

Theorem 2.4.15 *Let* $f^{*0} < aT$, $f^{*\infty} < aT$, *and assume there exists* $c_2 > 0$ *such that*

$$f(t, x) \geq aT\delta^2 c_2 \quad \text{for } x \in K \text{ and } c_2 \leq |x| \leq \delta c_2.$$

Then Eq. (2.33) has at least three positive T-periodic solutions for $\lambda = \frac{1}{T}$.

In particular, for $\lambda = 1$, the next theorem follows from Theorem 2.4.15.

Theorem 2.4.16 *Let* $f^{*0} < a$, $f^{*\infty} < a$, *and assume there exists a constant* $c_2 > 0$ *such that*

$$f(t, x) \geq a\delta^2 c_2 \quad \text{for } x \in K \text{ and } c_2 \leq |x| \leq \delta c_2.$$

Then Eq. (2.33) has at least three positive T-periodic solutions for $\lambda = 1$.

Theorem 2.4.17 *Let* $f^{*0} < aT$ *and* $f^{*\infty} < aT$. *If there exists* $c_2 > 0$ *such that*

$$f(t, x) \geq aT^2\delta^2 c_2 \quad \text{for } x \in K \text{ and } c_2 \leq |x| \leq \delta c_2,$$

then Eq. (2.33) has at least three positive T-periodic solutions for

$$\frac{e-1}{e^2 a T^2} \leq \lambda \leq \frac{e-1}{e a T^2}.$$

Theorem 2.4.18 *Let* $f^{*0} < a$, $f^{*\infty} < a$, *and assume there exists* $c_2 > 0$ *such that*

$$f(t, x) \geq aT\delta^2 c_2 \quad \text{for } x \in K \text{ and } c_2 \leq |x| \leq \delta c_2.$$

Then Eq. (2.33) has at least three positive T-periodic solutions for

$$\frac{e-1}{e^2 a T} \leq \lambda \leq \frac{e-1}{e a T}.$$

2.5 Applications to Some Mathematical Models

In this section, we apply some of the results obtained in Sect. 2.1 to models of the form of (2.10)–(2.12). In what follows, all the parameters in models (2.10)–(2.12) are assumed to be positive constants.

Example 2.5.1 If $a(t) \equiv a, b(t) \equiv b, \tau(t) \equiv \tau$, and $\gamma(t) \equiv \gamma$ are positive constants, then (2.10) reduces to

$$x'(t) = -ax(t) + be^{-\gamma x(t-\tau)}. \tag{2.34}$$

Graef et al. [2] and Zhang et al. [10] proved that (2.34) has at least one positive periodic solution. However, to the best of our knowledge, there is no such result for the existence of at least three positive periodic solutions of (2.34). It would be of interest to obtain such results. The following result follows from Theorem 2.1.7.

Theorem 2.5.1 *Let* $\gamma < 2e, \delta \leq \frac{2e}{2e-\gamma}$, *and* $\gamma\delta^2 < \delta - 1$ *hold, where* $\delta = e^{aT}$. *Then* (2.34) *has at least three positive T-periodic solutions for* $\frac{1}{2T} < b < \frac{1}{T}$.

Proof Let $f(t, x) = e^{-\gamma x}$. Then $f(t, x) > e^{-\gamma\delta c_2}$ for $c_2 \leq x \leq \delta c_2$, where $\delta = e^{aT}$. Thus, (H_8) holds if and only if $e^{-\gamma\delta c_2} \geq 2(\delta - 1)c_2$ for $c_2 \leq x \leq \delta c_2$. Now choose $c_2 = \frac{1}{\delta\gamma}$. Then $\delta \leq \frac{2e}{2e-\gamma}$ and $c_2 = \frac{1}{\delta\gamma}$ imply that $e^{-\gamma\delta c_2} \geq 2(\delta - 1)c_2$ for $c_2 \leq x \leq \delta c_2$. Hence (H_8) is satisfied. It is clear that $\tilde{f}^\infty < T$. In order to apply Theorem 2.1.7, we need to show the existence of a constant c_1 such that $0 < c_1 < c_2$ and (H_4) holds. Since $f(t, x) < 1$, (H_4) holds if $c_1 > \frac{\delta}{\delta-1}$. Indeed, (H_4) holds if $1 < \frac{\delta-1}{\delta}c_1$ for $0 \leq x \leq c_1$, that is, $c_1 > \frac{\delta}{\delta-1}$. Now we show the existence of c_1. Clearly, $\gamma\delta^2 < \delta - 1$ implies that $\frac{\delta}{\delta-1} < \frac{1}{\gamma\delta} = c_2$. Thus, there exists a real $c_1 \in (\frac{\delta}{\delta-1}, \frac{1}{\gamma\delta})$ such that $\frac{\delta}{\delta-1} < c_1 < c_2 = \frac{1}{\gamma\delta}$, so $f(t, x)$ satisfies (H_4). Hence, by Theorem 2.1.7, Eq. (2.34) has at least three positive T-periodic solutions for $\frac{1}{2T} < b < \frac{1}{T}$. This proves the theorem. $\qquad\square$

Example 2.5.2 If $a(t) \equiv a, b(t) \equiv b, \tau(t) \equiv \tau$, and $\gamma(t) \equiv \gamma$ are positive constants, then (2.11) reduces to

$$x'(t) = -ax(t) + bx^m(t - \tau)e^{-\gamma x^n(t-\tau)}. \tag{2.35}$$

Theorem 2.5.2 *Let* $m > 1$ *and* $2e(\delta - 1)\delta^{(m-1)}\gamma^{\frac{m-1}{n}} \leq 1$. *Then Eq.* (2.35) *has at least three positive T-periodic solutions for*

$$\frac{1}{2T} < b < \frac{1}{T}.$$

Proof Let $f(t, x) = x^m e^{-\gamma x^n}$ and set $c_2 = \frac{1}{\delta\gamma^{1/n}}$. Then it is easy to observe that $c_3 = \frac{1}{\gamma^{1/n}}$, and $2e(\delta - 1)\delta^{(m-1)}\gamma^{\frac{m-1}{n}} \leq 1$ imply that $c_2^m e^{-\gamma\delta^n c_2^n} > 2(\delta - 1)c_2$ for $c_2 \leq x \leq \delta c_2$. Hence, (H_8) is satisfied. Moreover, $\tilde{f}^\infty = 0 < T$ and $\tilde{f}^0 = 0 <$

T hold. Then, by Theorem 2.1.9, Eq. (2.35) has at least three positive T-periodic solutions for $\frac{1}{2T} < b < \frac{1}{T}$. $\square$

Although, the condition in Theorem 2.5.2 looks complicated, it is easy to verify. The following corollary follows from Theorem 2.5.2.

Corollary 2.5.1 *Let* $m > 1$ *and* $\delta < \min\left\{\frac{1}{\gamma^{1/n}}, \frac{1+2e}{2e}\right\}$. *Then* (2.35) *has at least three positive* T-*periodic solutions for* $\frac{1}{2T} < b < \frac{1}{T}$.

Proof In fact, $\delta < \min\left\{\frac{1}{\gamma^{1/n}}, \frac{1+2e}{2e}\right\}$ implies that $2e(\delta - 1)\delta^{(m-1)}\gamma^{\frac{m-1}{n}} \leq 1$ and hence by Theorem 2.5.2, (2.35) has at least three positive T-periodic solutions for $\frac{1}{2T} < b < \frac{1}{T}$. This completes the proof. $\square$

Example 2.5.3 If $a(t) \equiv a$, $b(t) \equiv b$, and $\tau(t) \equiv \tau$, are positive constants, and $m = 1$, then (2.12) reduces to

$$x'(t) = -ax(t) + b\frac{x(t - \tau)}{1 + x^n(t - \tau)}. \tag{2.36}$$

Applying Theorem 2.1.9 to Eq. (2.36) we have the following result.

Theorem 2.5.3 *Let* $e^{aT} < \frac{3}{2}$ *and* $T > 1$. *Then* (2.36) *has at least three positive* T-*periodic solutions for*

$$\frac{1}{2T} < b < \frac{1}{T}.$$

Proof Let $f(t, x) = \frac{x}{1+x^n}$. Then $\tilde{f}^\infty < T$ and $\tilde{f}^0 = 1 < T$. Choose $c_2 = \frac{1}{\delta}\left[\frac{1}{2(\delta-1)} - 1\right]^{\frac{1}{n}}$. Since $e^{aT} < \frac{3}{2}$, that is, $\delta < 3/2$, we have $c_2 > 0$. In addition, $c_2 = \frac{1}{\delta}\left[\frac{1}{2(\delta-1)} - 1\right]^{\frac{1}{n}}$ implies that $\frac{1}{1+\delta^n c_2^n} = 2(\delta - 1)$ and hence (H_8) holds. Then, by Theorem 2.1.9, (2.36) has at least three positive T-periodic solutions for

$$\frac{1}{2T} < b < \frac{1}{T}. \tag{}$$
     $\square$

Similarly, applying Theorem 2.1.9 to the autonomous equation

$$x'(t) = -ax(t) + b\frac{x(t - \tau)}{r + x^n(t - \tau)}, \tag{2.37}$$

we obtain the following theorem.

Theorem 2.5.4 *Let* $rT > 1$ *and* $e^{aT} < \frac{3}{2}$. *Then* (2.37) *has at least three positive* T-*periodic solutions for*

$$\frac{a}{2T} < b < \frac{a}{T},$$

where $\tau > 0$ is a constant.

Example 2.5.4 If $a(t) \equiv a, b(t) \equiv b$ and $\tau(t) \equiv \tau$ are constants, then (2.12) reduces to

$$x'(t) = -ax(t) + b\frac{x^m(t - \tau)}{1 + x^n(t - \tau)}. \tag{2.38}$$

Set

$$\mu = 2(\delta - 1)\delta^{2m-1}\frac{n}{1 + n - m}\left(\frac{1 + n - m}{m - 1}\right)^{\frac{m-1}{n}}. \tag{2.39}$$

Applying Theorem 2.1.9 to Eq. (2.38), we obtain the following result.

Theorem 2.5.5 *Let $0 < m - 1 < n$. Eq. (2.38) has at least three positive T-periodic solutions for $\frac{\mu}{2T} < b < \frac{\mu}{T}$, where μ is given in* (2.39).

Proof Now $\delta > 1$ and $0 < m - 1 < n$ implies $\mu > 0$. Eq. (2.38) can be written as

$$x'(t) = -ax(t) + \frac{b}{\mu}\mu\frac{x^m(t - \tau)}{1 + x^n(t - \tau)}. \tag{2.40}$$

Let $f(t, x) = \mu\frac{x^m}{1+x^n}$. Since $m > 1$, $\tilde{f}^0 = 0 < T$ and $\tilde{f}^\infty = 0 < T$. To complete the proof of the theorem, in view of Theorem 2.1.9, we need to find $c_2 > 0$ such that (H_8) holds. Set $c_2 = \frac{1}{\delta}\left(\frac{m-1}{1+n-m}\right)^{\frac{1}{n}}$. Now, for $c_2 \leq \|x\| \leq \delta c_2$, we have

$$\mu\frac{x^m}{1 + x^n} \geq \mu\frac{(\|x\|/\delta)^m}{1 + \delta^n c_2^n} \geq \frac{\mu}{\delta^m}\frac{c_2^m}{1 + \delta^n c_2^n} \tag{2.41}$$

and $1 + \delta^n c_2^n = \frac{n}{1+n-m}$. Then, from (2.41) and (2.39),

$$\mu\frac{x^m}{1 + x^n} \geq \frac{c_2^m}{\delta^m}\frac{n - m + 1}{n}2(\delta - 1)\delta^{2m-1}\frac{n}{n - m + 1}\left(\frac{1 + n - m}{m - 1}\right)^{\frac{m-1}{n}}$$

$$\geq 2(\delta - 1)c_2^m\delta^{m-1}\left(\frac{1 + n - m}{m - 1}\right)^{\frac{m-1}{n}}$$

$$\geq 2(\delta - 1)c_2^m\delta^{m-1}\frac{1}{\delta^{m-1}c_2^{m-1}}$$

$$\geq 2(\delta - 1)c_2.$$

This completes the proof of the theorem.   □

References

1. Deimling, K.: Nonlinear Functional Analysis. Springer, Berlin (1985)
2. Graef, J.R., Qian, C., Spikes, P.W.: Oscillation and global attractivity in a periodic delay equation. Can. Math. Bull. **38**, 275–283 (1996)
3. Han, F., Wang, Q.: Existence of multiple positive periodic solutions for differential equation with state-dependent delays. J. Math. Anal. Appl. **324**, 908–920 (2006)
4. Krasnosel'skii, M.A.: Positive Solution of Operator Equations. Noordhoff, Groningen (1964)
5. Padhi, S., Srivastava, S.: Existence of three periodic solutions for a nonlinear first order functional differential equation. J. Franklin Inst. **346**, 818–829 (2009)
6. Padhi, S., Qian, C., Srivastava, S.: Multiple periodic solutions for a first order nonlinear functional differential equation with applications to population dynamics. Comm. Appl. Anal. **12**, 341–352 (2008)
7. Padhi, S., Srivastava, S., Pati, S.: Three periodic solutions for a nonlinear first order functional differential equation. Appl. Math. Comput. **216**, 2450–2456 (2010)
8. Padhi, S., Srivsatava, S., Pati, S.: Positive periodic solutions for first order functional differential equations. Comm. Appl. Anal. **14**, 447–462 (2010)
9. Royden, H.L.: Real Analysis. Prentice Hall of India, New Delhi (1995)
10. Zhang, W., Zhu, D., Bi, P.: Existence of periodic solutions of a scalar functional differential equation via a fixed point theorem. Math. Comput. Model. **46**, 718–729 (2007)

Chapter 3
Multiple Periodic Solutions of a System of Functional Differential Equations

This chapter[1] deals with the existence of three positive periodic solutions with positive components, to the system of differential equations

$$x'(t) = A(t, x)x(t) + \lambda f(t, x_t). \tag{3.1}$$

Here, A is a diagonal $n \times n$ matrix whose entries depend on t and on the unknown function $x = (x_1, x_2, \ldots, x_n)^T$. We assume that the diagonal entries $a_i(t, x) \in C(R \times R, R)$ are periodic in t, with a common period T. The parameter λ is positive and assumed to be known, or at least its range is known. The function $f = (f_1, f_2, \ldots, f_n)^T$ has each component in $C(R \times R, R)$ and is T-periodic in the first variable. Here, x_t denotes a functional that depends on t and satisfies the conditions stated below. Typical examples of such functionals are evaluations at $x(h(t))$ and memory terms such as $\int_{-\infty}^{t} k(s)x(s)\,\mathrm{d}s$.

The above system includes the scalar differential equations

$$x'(t) = a(t)g(x(t))x(t) - \lambda b(t) f(t, x(t - \tau)), \tag{3.2}$$

$$x'(t) = \mp a(t)x(t) \pm \lambda f(t, x(t - \tau)) \tag{3.3}$$

that have been studied in Chap. 2 and by several authors [1, 3–8, 10–14]. Jiang et al. [5] proved the existence and nonexistence of a nonnegative periodic solution of (3.1) if $A(t, x) = A(t)$. Zhang et al. [14] used a fixed point theorem in cone expansion and cone compression [2] to show the existence of two periodic solutions for the above equation.

All functions in this chapter are assumed to be in the space X of positive T-periodic continuous functions, equipped with the supremum norm $\|x\| = \max_{1 \le i \le n} \sup_t |x_i(t)|$. We use the following as general assumptions throughout this chapter.

---

[1] Some of the results in this chapter are taken from Padhi et al. [9].

S. Padhi et al., *Periodic Solutions of First-Order Functional Differential Equations in Population Dynamics*, DOI: 10.1007/978-81-322-1895-1_3, © Springer India 2014

(A1) There exist continuous T-periodic functions b and c such that $0 \leq b(t) \leq |a_i(t, x)| \leq c(t)$ for $1 \leq i \leq n$ and all T-periodic functions x. Furthermore, $\int_0^T b(t)\,dt > 0$.

(A2) $f_i(t, x_t) \int_0^T a_i(s, x)\,ds \leq 0$ for $1 \leq i \leq n$ and $0 \leq t \leq T$.

(A3) $f(t, x)$ is a continuous function of x.

(A4) For any $L > 0$ and $\varepsilon > 0$, there exists $\delta > 0$ such that if $\phi, \psi \in X$ with $\|\phi\| \leq L$ $\|\psi\| \leq L$, and $\|\phi - \psi\| < \delta$, we have

$$|f(t, \phi) - f(t, \psi)| < \varepsilon$$

uniformly in t.

We can construct the Green's kernel so that solutions of (3.1) satisfy the integral equation

$$x(t) = \lambda \int_t^{t+T} G(t, s) f(s, x_s)\,ds,$$

where $G(t, s)$ is a diagonal matrix with entries

$$G_i(t, s) = \frac{\exp\left(\int_s^t a_i(\theta, x_\theta)\,d\theta\right)}{\exp\left(-\int_0^T a_i(\theta, x_\theta)\,d\theta\right) - 1}.$$

These entries are bounded as follows:

$$\alpha = \frac{\exp\left(-\int_0^T c(\theta)\,d\theta\right)}{\exp\left(-\int_0^T b(\theta)\,d\theta\right) - 1} \leq |G_i(t, s)| \leq \frac{\exp\left(\int_0^T c(\theta)\,d\theta\right)}{\exp\left(-\int_0^T c(\theta)\,d\theta\right) - 1} = \beta. \quad (3.4)$$

Note that $G(t, s)$ is T-periodic in both variables and that G_i and $\int_0^T a_i$ have opposite signs. Therefore, by (A2), f_i and G_i have the same sign.

We define the operator A_λ on X by

$$(A_\lambda x)(t) = \lambda \int_t^{t+T} G(t, s) f(s, x_s)\,ds.$$

Since the functions G_i and f_i have the same sign, $A_\lambda x_i$ is nonnegative. Furthermore,

$$(A_\lambda x_i)(t) = \lambda \int_t^{t+T} |G_i(t, s)| |f_i(s, x_s)|\,ds \leq \lambda \beta \int_0^T |f_i(s, x_s)|\,ds.$$

Taking the supremum on t,

$$\|(A_\lambda x_i)(t)\| \le \lambda \beta \int_0^T |f_i(s, x_s)| \, ds.$$

Also, we have

$$(A_\lambda x_i)(t) \ge \lambda \alpha \int_0^T |f_i(s, x_s)| \, ds.$$

Combining the two inequalities above, we obtain

$$(A_\lambda x_i)(t) \ge \frac{\alpha}{\beta} \|(A_\lambda x_i)(t)\|. \tag{3.5}$$

Motivated by this inequality, we define the cone K in X as

$$K = \{x \in X : x_i(t) \ge \frac{\alpha}{\beta} \|x_i\| \text{ for } t \in [0, T] \text{ and } 1 \le i \le n\}.$$

Similar to what we did in Lemma 2.1.1, we can show that $A_\lambda(K) \subset K$ and A_λ is compact and completely continuous.

3.1 Positive Periodic Solutions of the Equation $x'(t) = A(t, x)x(t) + \lambda f(t, x_t)$

In this section, sufficient conditions are obtained for the existence of positive periodic solutions of the system of functional differential equation (3.1).

To prove the main result, we state the following conditions in terms of the bounds α and β defined by (3.4):

(H_{20}) There exists a positive constant c_1 such that

$$\lambda \beta \int_0^T |f_i(t, x_t)| dt < c_1 \quad \text{for } 1 \le i \le n \quad \text{and} \quad x \in K \quad \text{with} \quad \|x\| \le c_1;$$

(H_{21}) There exists a positive constant $c_2 > c_1$ such that

$$c_2 < \lambda \alpha \int_0^T |f_i(t, x_t)| dt \quad \text{for } 1 \le i \le n \quad \text{and } x \in K \quad \text{with } c_2 \le \|x\| \le \beta c_2 \alpha;$$

and

(H_{22}) There exists a constant $c_4 \geq \beta c_2/\alpha = c_3$ such that

$$\lambda \beta \int_0^T |f_i(t, x_t)| dt \leq c_4 \quad \text{for} \quad 1 \leq i \leq n \quad \text{and} \quad x \in K \quad \text{with} \quad \|x\| \leq c_4.$$

Theorem 3.1.1 *Under the conditions (H_{20})–(H_{22}), the hypotheses of the Leggett– Williams fixed point Theorem 1.2.2 are satisfied and therefore Eq.(3.1) has at least three positive T-periodic solutions.*

Proof First, we show that A_λ maps $\overline{K}_{c_4}$ into $\overline{K}_{c_4}$. For x in $\overline{K}_{c_4}$, using (3.4) and (H_{22}), we have

$$\|A_\lambda x_i\| \leq \lambda \int_t^{t+T} |G_i(t, s)||f_i(s, x_s)| ds$$

$$\leq \lambda \beta \int_0^T |f_i(s, x_s)| ds$$

$$\leq c_4,$$

which proves that $A_\lambda(\overline{K}_{c_4}) \subset \overline{K}_{c_4}$.

Now, we prove (i) in Theorem 1.2.2. We define a nonnegative concave continuous functional

$$\psi(x) = \min_{1 \leq i \leq n} \inf_{0 \leq t \leq T} x_i(t).$$

The set $\{x \in K(\psi, c_2, c_3) : \psi(x) > c_2\}$ is not empty because the constant function $x_i(t) = (c_2 + c_3)/2$ is in this set. For $x \in K(\psi, c_2, c_3)$, we have $c_2 \leq \|x\| \leq c_3 = \beta c_2/\alpha$. Then, using the definition of ψ, the fact that G_i and f_i have the same signs, the bounds (3.4), and condition (H_{21}), we have

$$\psi(A_\lambda x) = \min_i \inf_t \lambda \int_t^{t+T} |G_i(t, s)||f_i(s, x_s)| ds$$

$$\geq \min_i \lambda \alpha \int_0^T |f_i(s, x_s)| ds$$

$$> c_2,$$

which implies (i) in Theorem 1.2.2 holds. Now, for x in $\overline{K}_{c_1}$, using (3.4) and (H_{20}), we have

$$\|A_\lambda x_i\| \leq \lambda \int_t^{t+T} |G_i(t, s)||f_i(s, x_s)|\, ds$$

$$\leq \lambda\beta \int_0^T |f_i(s, x_s)|\, ds$$

$$< c_1,$$

which shows (ii) in Theorem 1.2.2 holds. Finally, from the definition of ψ and (3.4),

$$\psi(A_\lambda x) = \min_i \inf_t \lambda \int_t^{t+T} |G_i(t, s)||f_i(s, x_s)|\, ds$$

$$\geq \min_i \lambda\alpha \int_0^T |f_i(s, x_s)|\, ds.$$

From $\|A_\lambda x_i\| > c_3 = \dfrac{\beta}{\alpha} c_2$ and (3.4), we have

$$c_3 < \|A_\lambda x_i\| \leq \lambda\beta \int_0^T |f_i(s, x_s)|\, ds.$$

The above two inequalities give $\psi(A_\lambda x) > c_2$, which proves (iii) in Theorem 1.2.2. Hence, all the conditions of the Leggett–Williams fixed point Theorem 1.2.2 are satisfied, so the operator A_λ has at least three fixed points that correspond to periodic solutions of (3.1). This proves the theorem. $\qquad\square$

Other versions of Theorem 3.1.1 can be stated using the following four lemmas.

Lemma 3.1.1 *Condition (H_{20}) is implied by*

$$\beta\lambda \limsup_{\|x\|\to 0} \int_0^T |f_i(t, x_t)|\, dt/\|x\| < 1 \quad \text{for all} \quad i.$$

In turn, this condition is implied by

$$\beta\lambda T \limsup_{\|x\|\to 0} \max_{0\leq t\leq T} |f_i(t, x_t)|/\|x\| < 1 \quad \text{for all} \quad i.$$

Proof From the definition of limit superior, for each $0 < \epsilon < 1$, there exists a $\delta > 0$ such that

$$\beta\lambda \int_0^T |f_i(t, x_t)|\, dt < \epsilon \|x\| \quad \text{for} \quad \|x\| < \delta \quad \text{for all} \quad i.$$

Select $c_1 < \delta$. Then, for x in $\overline{K}_{c_1}$,

$$\beta\lambda \int_0^T |f_i(t, x_t)|\, dt < \epsilon \|x\| < \|x\| \leq c_1 \quad \text{for all} \quad i$$

which implies (H_{20}). □

Lemma 3.1.2 *Condition (H_{21}) is implied by*

$$c_2 < \lambda\alpha T |f_i(t, x_t)| \quad \text{for} \quad t \in [0, T], \ \forall i, \ x \in K \quad \text{with} \quad c_2 \leq \|x\| \leq \frac{\beta}{\alpha} c_2.$$

Lemma 3.1.3 *Condition (H_{22}) is implied by*

$$\beta\lambda \limsup_{\|x\|\to\infty} \int_0^T |f_i(t, x_t)|\, dt / \|x\| < 1 \quad \text{for all} \quad i.$$

In turn, this condition is implied by

$$\beta\lambda T \limsup_{\|x\|\to\infty} \max_{0\leq t\leq T} |f_i(t, x_t)| / \|x\| < 1 \quad \text{for all} \quad i.$$

Proof Clearly, there exist positive constants $\epsilon < 1$ and $\delta > 0$ such that

$$\beta\lambda \int_0^T |f_i(t, x_t)|\, dt < \epsilon \|x\| \quad \text{for} \quad \|x\| \geq \delta \quad \text{for all} \quad i.$$

From (A3), if $\|x\| \leq \delta$, $\beta\lambda \int_0^T |f_i(t, x_t)|\, dt$ is bounded by a positive constant r. Select $c_4 = \dfrac{r}{1-\epsilon}$. Then, for $x \in \overline{K}_{c_4}$,

$$\beta\lambda \int_0^T |f_i(t, x_t)|\, dt \leq \epsilon \|x\| + r$$

$$\leq \epsilon c_4 + r$$

$$= c_4,$$

which implies (H_{22}). □

When we know only the range of λ, rather than its value, hypotheses (H_{20})–(H_{22}) need to be modified as follows.

Theorem 3.1.2 *When* $\lambda_1 \leq \lambda \leq \lambda_2$, *Theorem* 3.1.1 *remains valid if we replace* (H_{20})–(H_{22}), *respectively, by*

(H_{23}) $\lambda_2 \beta \int_0^T |f_i(t, x_t)| dt < c_1$ *for* $1 \leq i \leq n$ *and* $x \in K$ *with* $\|x\| \leq c_1$,

(H_{24}) $c_2 < \lambda_1 \alpha \int_0^T |f_i(t, x_t)| dt$ *for* $1 \leq i \leq n$ *and* $x \in K$ *with* $c_2 \leq \|x\| \leq \frac{\beta}{\alpha} c_2$

and

(H_{25}) $\lambda_2 \beta \int_0^T |f_i(t, x_t)| dt \leq c_4$ *for* $1 \leq i \leq n$ *and* $x \in K$ *with* $\|x\| \leq c_4$.

3.2 Applications to Some Mathematical Models

Example 3.2.1 As a particular case of (3.1), we consider the scalar Eq. (2.12), which is a Hematopoiesis model. Here, we assume that a, b and τ are continuous periodic positive functions with a common period T, and the constants m, n, and T are positive. The Green's kernel for this equation is

$$G(t, s) = \frac{\exp\left(\int_t^s a(\theta)\, d\theta\right)}{\delta - 1}, \quad \text{where} \quad \delta = \exp\left(\int_0^T a(\theta)\, d\theta\right).$$

This kernel has the bounds

$$\frac{1}{\delta - 1} \leq G(t, s) \leq \frac{\delta}{\delta - 1}.$$

We define the cone K on X by

$$K = \{x \in X : x(t) \geq \frac{1}{\delta}\|x\|, \quad \text{for} \quad t \in [0, T]\},$$

and the operator A by

$$(Ax)(t) = \int_t^{t+T} G(t, s) b(s) \frac{x^m(s - \tau(s))}{1 + x^n(s - \tau(s))}\, ds.$$

As we see in the proof of the next theorem, condition (i) of the Leggett–Williams fixed point Theorem 1.2.2 is satisfied if there exists a positive constant c_2 such that

$$\frac{1}{\delta - 1} \int\limits_{t}^{t+T} b(s) \frac{(c_2/\delta)^m}{1 + (\delta c_2)^n} \, ds > c_2.$$

We select c_2 as the minimizer of the function $f(c) = (\delta - 1)c(\frac{c}{\delta})^{-m}(1 + (\delta c)^n)$. This choice of c_2 leads the assumption (3.6) in the following result.

Theorem 3.2.1 *Assume $n > m - 1 > 0$ and*

$$\int\limits_{0}^{T} b(s) \, ds > \delta^{2m-1}(\delta - 1) \left(\frac{n}{1 + n - m} \right) \left(\frac{1 + n - m}{m - 1} \right)^{\frac{m-1}{n}}. \qquad (3.6)$$

Then, the hypotheses of Theorem 1.2.2 *are satisfied and therefore* (2.12) *has at least three positive T-periodic solutions.*

Remark 3.2.1 Equation (2.12) with constant coefficients was considered in Sect. 2.4 of Chap. 2 (see Example 2.5.4 and Theorem 2.5.5). If $a(t) \equiv a$, $b(t) \equiv b$, and $\tau(t) \equiv \tau$ are constants, then Theorem 3.2.1 implies that, if $bT > \mu$, then (2.38) has at least three positive T-periodic solutions, where μ is given in (2.39). Here, only the lower bound on bT is required. On the other hand, a range on bT, that is, $\frac{\mu}{2} < bT < \mu$ is required in Theorem 2.5.5 for the existence of at least three positive T-periodic solutions of (2.38) although the condition $0 < m - 1 < n$ is required in both the Theorems 2.5.5 and 3.2.1. Thus, Theorem 3.2.1 extends Theorem 2.5.5.

Furthermore, if $a(t) \equiv a$, $b(t) \equiv b$, and $\tau(t) \equiv \tau$ are positive constants, then (3.6) gives a better lower bound on bT than (2.39) for the existence of at least three positive T-periodic solutions of (2.38). In view of this, one may observe, from the proof of Theorem 3.2.1, which is given below, that a direct application of the Leggett–Williams multiple fixed point theorem, that is, Theorem 1.2.2 provide a better sufficient condition for the existence of positive T-periodic solutions of models arising in ecological and biological systems.

It seems that no result exists on the multiple positive periodic solutions of (2.38) when $bT \leq \frac{\mu}{2}$. Note that, if $bT > \frac{\mu}{2}$, then (2.38) has at least three positive T-periodic solutions, by Theorem 3.2.1.

Proof of Theorem 3.2.1 From the definition of the cone K, $\|x\|/\delta \leq x(s - \tau(s)) \leq \|x\|$ for $s \in [0, T]$. Then,

$$\frac{1}{\|x\|} \int\limits_{0}^{T} G(t, s)b(s) \frac{x^m(s - \tau(s))}{1 + x^n(s - \tau(s))} \, ds \leq \frac{\delta}{\delta - 1} \int\limits_{0}^{T} b(s) \, ds \frac{\|x\|^{m-1}}{1 + (\|x\|/\delta)^n}.$$

Since $\lim\limits_{x \to \infty} x^{m-1}/(1 + x^n) = 0$, the right-hand side of the above inequality approaches zero as $\|x\| \to \infty$. By Lemma 3.1.3, this implies (H_{22}), for c_4 arbitrarily large, which in turn implies $A(\overline{K}_{c_4}) \subset \overline{K}_{c_4}$. On the other hand, since $\lim\limits_{x \to 0} x^{m-1}/(1+$ $x^n) = 0$, the right-hand side of the above inequality approaches zero as $\|x\| \to 0$. By Lemma 3.1.1, this implies (H_{20}), for c_1 arbitrarily small, which in turn implies (ii) of Theorem 1.2.2. To prove (i) of Theorem 1.2.2, we set

$$c_2 = \frac{1}{\delta} \left(\frac{m-1}{1+n-m} \right)^{1/n}$$

and $c_3 = \delta c_2$. Note that the set $\{x \in K(\psi, c_2, c_3) : \psi(x) > c_2\}$ is not empty because the constant function $\frac{c_2 + c_3}{2}$ is in this set. For each x in the cone K with $c_2 \le \psi(x)$ and $\|x\| \le c_3$, we have $c_2 \le \|x\| \le c_3$ and $\frac{c_2}{\delta} \le x(s - \tau(s)) \le c_2 \delta$. Then,

$$\psi(Ax) \ge \frac{1}{\delta-1} \int_0^T b(s)\,ds \frac{(c_2/\delta)^m}{1+(c_2\delta)^n} > c_2.$$

The above inequality follows from (3.6) and the choice of c_2. Here, we have used that $1 + (c_2\delta)^n = \frac{n}{1+n-m}$ and

$$\left(\frac{c_2}{\delta} \right)^m = c_2 \frac{1}{\delta^{2m-1}} \left(\frac{m-1}{1+n-m} \right)^{\frac{m-1}{n}}.$$

The above inequality implies (i) of Theorem 1.2.2. Now, we prove (iii) in Theorem 1.2.2 holds. Note that

$$\psi(Ax) \ge \frac{1}{\delta-1} \int_t^{t+T} b(s) \frac{x^m(s - \tau(s))}{1+x^n(s - \tau(s))}\,ds.$$

From $\|Ax\| > c_3 = \delta c_2$, we have

$$c_3 < \|Ax\| \le \frac{\delta}{\delta-1} \int_t^{t+T} b(s) \frac{x^m(s - \tau(s))}{1+x^n(s - \tau(s))}\,ds.$$

Combining the two inequalities above gives $\psi(Ax) > c_2$, which shows (iii) of Theorem 1.2.2 holds. By the Leggett–Williams Theorem 1.2.2, Eq. (2.12) has at least three positive T-periodic solutions. $\qquad\square$

Example 3.2.2 Consider the scalar delay differential equation

$$x'(t) = -a(t)x(t) + b(t)x^m(t - \tau(t))e^{-\gamma x(t-\tau(t))}, \tag{3.7}$$

where a, b, and τ are continuous periodic positive functions with a common period T, and the constants m, γ, and T are positive. The Green's kernel G and the cone K are the same as above, and the operator A is

$$(Ax)(t) = \int_t^{t+T} G(t, s)b(s)x^m(s - \tau(s)) \exp(-\gamma x(s - \tau(s))) \, ds.$$

For proving (i) in the Leggett–Williams fixed point Theorem 1.2.2, we need a positive constant c_2 such that

$$\frac{1}{\delta - 1} \int_t^{t+T} b(s) \, ds \left(\frac{c_2}{\delta}\right)^m \exp(-\gamma \delta c_2) > c_2.$$

We select c_2 as the minimizer of the function $f(c) = (\delta - 1)c \left(\frac{c}{\delta}\right)^{-m} e^{\gamma \delta c}$. This choice of c_2 leads to the assumption in the following result.

Theorem 3.2.2 *Assume that $m > 1$ and that*

$$\int_0^T b(s) \, ds > \delta(\delta - 1) \left(\frac{\gamma \delta^2 e}{m - 1}\right)^{m-1}. \tag{3.8}$$

Then, the hypotheses of Theorem 1.2.2 *are satisfied and therefore* (3.7) *has at least three positive T-periodic solutions.*

Remark 3.2.2 Clearly, (3.7) is a particular case of (2.11) with $n = 1$. If $a(t) \equiv a$, $b(t) \equiv b$, and $\tau(t) \equiv \tau$ are positive constants, then Theorem 3.2.2 implies that if $m > 1$ and

$$bT > \delta(\delta - 1) \left(\frac{\gamma \delta^2 e}{m - 1}\right)^{m-1}, \tag{3.9}$$

then Eq. (2.35) with $n = 1$ has at least three positive T-periodic solutions, where $\delta = e^{aT}$. On the other hand, Theorem 2.5.2 and Corollary 2.5.1 imply that if $m > 1$ and either

$$2e(\delta - 1)\delta^{m-1}\gamma^{m-1} \leq 1 \tag{3.10}$$

or

$$\delta < \min \left\{\frac{1}{\gamma}, \frac{1 + 2e}{2e}\right\}, \tag{3.11}$$

then Eq. (2.35) with $n = 1$ has at least three positive T-periodic solutions for $\frac{1}{2} < bT < 1$. Although the condition in (3.9) looks complicated, only a lower bound on bT is required. On the other hand, condition (3.10) or (3.11) seems easy to verify with the price that a lower and upper bound on bT are required.

Proof of Theorem 3.2.2 This proof is similar to the one in Theorem 3.2.1, so we only give a sketch of it. Since $\lim\limits_{x \to \infty} x^{m-1} e^{-\gamma x} = 0$, Lemma 3.1.3 implies (H_{22}), for c_4 arbitrarily large, which in turn implies $A(\overline{K}_{c_4}) \subset \overline{K}_{c_4}$. Again, since $\lim\limits_{x \to 0} x^{m-1} e^{-\gamma x} = 0$, Lemma 3.1.2 implies (H_{20}), for c_1 arbitrarily small, which in turn implies (ii) of Theorem 1.2.2. To prove (i) of Theorem 1.2.2, we set

$$c_2 = \frac{m-1}{\gamma \delta}$$

and $c_3 = \delta c_2$. For each x in the cone K with $c_2 \leq \psi(x)$ and $\|x\| \leq c_3$, we have $c_2 \leq \|x\| \leq c_3$ and $\frac{c_2}{\delta} \leq x(s - \tau(s)) \leq c_2 \delta$. Hence,

$$\psi(Ax) \geq \frac{1}{\delta - 1} \int_0^T b(s)\, ds \left(\frac{c_2}{\delta}\right)^m e^{-\gamma c_2 \delta} > c_2.$$

The above inequality follows from (3.8) and the choice of c_2. Therefore, condition (i) is satisfied. The proof of condition (iii) is the same as in the proof of Theorem 3.2.1. Then, by the Leggett–Williams fixed point Theorem 1.2.2, (3.7) has at least three positive T-periodic solutions. □

References

1. Bai, D., Xu, Y.: Periodic solutions of first order functional differential equations with periodic deviations. Comput. Math. Appl. **53**, 1361–1366 (2007)
2. Deimling, K.: Nonlinear Functional Analysis. Springer, Berlin (1985)
3. Jiang, D., Wei, J.J.: Existence of positive periodic solutions for non-autonomous functional differential equations with delay. Chin. Ann. Math. **20A**, 715–720 (1999). (in Chinese)
4. Jiang, D., Wei, J., Jhang, B.: Positive periodic solutions of functional differential equations and population models. Electron. J. Differ. Equ. **2002**(71), 1–13 (2002)
5. Jiang, D., O'Regan, D., Agrawal, R.P., Xu, X.J.: On the number of positive periodic solutions of functional differential equations and population models. Math. Models Meth. Appl. Sci. **15**, 555–573 (2005)
6. Jin, Z.L., Wang, H.: A note on positive periodic solutions of delayed differential equations. Appl. Math. Lett. **23**, 581–584 (2010)
7. Liu, B.: Positive periodic solution for a nonautonomous delay differential equation. Acta Math. Appl. Sinica Engl. Ser. **19**, 307–316 (2003)
8. Lu, S.P., Ge, W.: On the existence of positive periodic solutions for neutral functional differential equation with multiple deviating arguments. Acta Math. Appl. Sinica Engl. Ser. **19**, 631–640 (2003)
9. Padhi, S., Srivastava, S., Dix, J.G.: Existence of three nonnegative periodic solutions for functional differential equations and applications to hematopoiesis. PanAmer. Math. J. **19**, 27–36 (2009)
10. Wan, A., Jiang, D.: Existence of positive periodic solutions for functional differential equations. Kyushu J. Math. **56**, 193–202 (2002)

11. Wang, H.: Positive periodic solutions of functional differential equations. J. Differ. Equ. **202**, 354–366 (2004)
12. Wu, Y.: Existence of positive periodic solutions for a functional differential equation with a parameter. Nonlinear Anal. **68**, 1954–1962 (2008)
13. Ye, D., Fan, M., Wang, H.: Periodic solutions for scalar functional differential equations. Nonlinear Anal. **62**, 1157–1181 (2005)
14. Zhang, W., Zhu, D., Bi, P.: Existence of periodic solutions of a scalar functional differential equation via a fixed point theorem. Math. Comput. Model. **46**, 718–729 (2007)

Chapter 4
Multiple Periodic Solutions of Nonlinear Functional Differential Equations

In this chapter[1], we present results on the existence of two positive periodic solutions of the first order functional differential equation

$$x'(t) = a(t)x(t) - f(t, x(h(t))), \tag{4.1}$$

where $a, h \in C(R, R_+)$ and $a(t + T) = a(t)$, $T > 0$ is a real number, $f : R \times R_+ \to R_+$, and $f(t + T, x) = f(t, x)$. If $h(t) = t - \tau(t)$ and $\tau \in C(R, R_+)$, $\tau(t + T) = \tau(t)$ with $\tau(t) \le t$, then (4.1) takes the form

$$x'(t) = a(t)x(t) - f(t, x(t - \tau(t))). \tag{4.2}$$

From results on the existence of positive periodic solutions of (4.1), we can find from the arguments in the succeeding sections that some similar results can be derived for (4.2). The results obtained in [1, 5, 7, 12–14] can be applied to (4.1). One may observe from the sufficient conditions assumed in Chaps. 2 and 3, that the function f needs to be unimodal, that is, the function f first increases and then it decreases eventually. This is because of the choice of a constant c_4 needed in the use of the Leggett-Williams multiple fixed point Theorem 1.2.2, for the existence of three fixed points of an operator which, in turn, is equivalent to the existence of three positive periodic solutions of (4.1) or (4.2). The above choices of functions exclude many important class of growth functions arising in various mathematical models, such as:

(i) Logistic equation of multiplicative type with several delays

$$x'(t) = x(t)\left[a(t) - \prod_{i=1}^{n} b_i(t)x(t - \tau_i(t))\right], \tag{4.3}$$

where $a, b_i, \tau_i \in C(R, R_+)$ are T-periodic functions;
(ii) Generalized Richards single species growth model

---

[1] Some of the results in this chapter are taken from Padhi et al. [9–11].

S. Padhi et al., *Periodic Solutions of First-Order Functional Differential Equations in Population Dynamics*, DOI: 10.1007/978-81-322-1895-1_4, © Springer India 2014

$$x'(t) = x(t)\left[a(t) - \left(\frac{x(t - \tau(t))}{E(t)}\right)^{\theta}\right], \tag{4.4}$$

where $a, E, \tau \in C(R, R_+)$ are T-periodic functions and $\theta > 0$ is a constant;
(iii) Generalized Michaelis-Menton type single species growth model

$$x'(t) = x(t)\left[a(t) - \sum_{i=1}^{n} \frac{b_i(t)x(t - \tau_i(t))}{1 + c_i(t)x(t - \tau_i(t))}\right], \tag{4.5}$$

where a, b_i, c_i, and $\tau_i \in C(R, R_+), i = 1, 2, \ldots, n$, are T-periodic functions.

In this chapter, we attempt to study the existence of two positive T-periodic solutions of the Eq. (4.1). Then we apply the obtained result to find sufficient conditions for the existence of two positive T-periodic solutions of the models (4.3)–(4.5). To prove the results, we use the Leggett-Williams multiple fixed point Theorem 1.2.1.

The following open problem was proposed by Kuang [6, open Problem 9.2]: *Obtain sufficient conditions for the existence of positive periodic solutions of the equation*

$$x'(t) = x(t)[a(t) - b(t)x(t) - c(t)x(t - \tau(t)) - d(t)x'(t - \sigma(t))]. \tag{4.6}$$

Liu et al. [8] gave a partial answer to the above problem by using a fixed point theorem for strict set-contractions. They proved that (4.6) has at least one positive T-periodic solution. Freedman and Wu [3] studied the existence and global attractivity of a positive periodic solution of (4.6) with $d(t) \equiv 0$. In this chapter, we apply the Leggett-Williams multiple fixed point Theorem 1.2.1 to show that (4.6) has at least two positive T-periodic solutions (See Example 4.2.1) when $d(t) \equiv 0$.

The results of this chapter can be extended to

$$x'(t) = a(t)x(t) - f(t, x(h_1(t)), \ldots, x(h_n(t))), \tag{4.7}$$

where $h_i(t) \geq 0, i = 1, \ldots, n$, and $f \in C(R \times R_+^n, R_+)$ is periodic with respect to the first variable.

Observe that (4.1) is equivalent to

$$x(t) = \int_t^{t+T} G(t, s) f(s, x(h(s))) \, ds,$$

where

$$G(t, s) = \frac{e^{-\int_t^s a(\theta) \, d\theta}}{1 - e^{-\int_0^T a(\theta) \, d\theta}}$$

is the Green's kernel. The lower bound, being positive, is used for defining a cone. The Green's kernel $G(t, s)$ satisfies the property

$$0 < \alpha = \frac{\delta}{1 - \delta} \le G(t, s) \le \frac{1}{1 - \delta} = \beta, \quad s \in [t, t + T],$$

where $\delta = e^{-\int_0^T a(\theta)\, d\theta} < 1$.

Let

$$X = \{x \in C(R, R) : x(t) = x(t + T)\}$$

with the norm $\|x\| = \sup_{t \in [0, T]} |x(t)|$; then X is a Banach space with the norm $\| \cdot \|$.
Define a cone K in X by

$$K = \{x \in X : x(t) \ge \delta \|x\|, \ t \in [0, T]\}$$

and an operator A on X by

$$(Ax)(t) = \int_t^{t+T} G(t, s) f(s, x(h(s)))\, ds. \tag{4.8}$$

If we proceed along the lines of Lemma 2.1.1 and Lemma 2.1.2 in Chap. 2, we can prove that $A(K) \subset K$, $A : K \to K$ is completely continuous, and the existence of a positive periodic solution of (4.1) is equivalent to the existence of a fixed point of A in K.

4.1 Positive Periodic Solutions of the Equation $x'(t) = a(t)x(t) - f(t, x(h(t)))$

In this section, we shall obtain some sufficient conditions for the existence of at least two positive T-periodic solutions of (4.1).
    Denote

$$f^\theta = \limsup_{x \to \theta} \frac{f(t, x)}{a(t)x} \quad \text{and} \quad F^\theta = \limsup_{x \to \theta} \frac{f(t, x)}{x}.$$

Theorem 4.1.1 *Assume that there exist constants c_1 and c_2 with $0 < c_1 < c_2$ such that*

$$(H_{26}) \quad \int_0^T f(t, x(h(t)))\, dt > \frac{c_2}{\alpha} \quad for \ x \in K, \ c_2 \le x \le \frac{c_2}{\delta}, \ and \ 0 \le t \le T,$$

and

$$(H_{27}) \quad \int_0^T f(t, x)) \, dt < \frac{c_1}{\beta} \; for \; x \in K, \; 0 \le x \le c_1, \; and \; 0 \le t \le T.$$

Then Eq. (4.1) has at least two positive T-periodic solutions.

Proof Define a nonnegative concave continuous functional ψ on K by $\psi(x) = \min_{t \in [0,T]} x(t)$. Then $\psi(x) \le \|x\|$. Set $c_3 = \frac{c_2}{\delta}$ and $\phi_0(t) = \phi_0 = \frac{c_2 + c_3}{2}$. Then $\phi_0 \in \{x \in K(\psi, c_2, c_3) : \psi(x) > c_2\}$. Furthermore, for $x \in K(\psi, c_2, c_3)$, (H_{26}) implies

$$\psi(Ax) = \min_{0 \le t \le T} \int_t^{t+T} G(t, s) f(s, x(h(s))) \, ds$$

$$\ge \alpha \int_0^T f(s, x(h(s))) \, ds$$

$$> c_2.$$

Now let $x \in \overline{K}_{c_1}$. Then, from (H_{27}),

$$\|Ax\| = \sup_{0 \le t \le T} \int_t^{t+T} G(t, s) f(s, x(h(s))) \, ds$$

$$\le \beta \int_0^T f(s, x(h(s))) \, ds$$

$$< c_1.$$

Next, suppose that $x \in \overline{K}_{c_3}$ with $\|Ax\| > c_3$. Then,

$$\psi(Ax) = \min_{0 \le t \le T} \int_t^{t+T} G(t, s) f(s, x(h(s))) \, ds$$

$$\ge \alpha \int_0^T f(s, x(h(s))) \, ds$$

and

$$c_3 < \|Ax\| \le \beta \int_0^T f(s, x(h(s))) \, ds$$

$$= \frac{\alpha}{\delta} \int_0^T f(s, x(h(s))) \, ds$$

$$\le \frac{1}{\delta} \psi(Ax)$$

imply that

$$\psi(Ax) > \frac{c_2}{c_3} \|Ax\|.$$

Hence, by Theorem 1.2.1, Eq. (4.1) has at least two positive T-periodic solutions. This completes the proof of the theorem. □

Theorem 4.1.2 *Assume that there exist constants c_1 and c_2 with $0 < c_1 < c_2$ such that*

(H_{28}) $f(t, x(h(t))) > \dfrac{c_2}{\alpha T}$ *for $x \in K$, $c_2 \le x \le \dfrac{c_2}{\delta}$, and $0 \le t \le T$*

and

(H_{29}) $f(t, x(h(t))) < \dfrac{c_1}{\beta T}$ *for $x \in K$, $0 \le x \le c_1$, and $0 \le t \le T$.*

Then Eq. (4.1) has at least two positive T-periodic solutions.

The proof of the theorem follows from Theorem 4.1.1. Indeed, (H_{26}) and (H_{27}) follow from (H_{28}) and (H_{29}), respectively.

Theorem 4.1.3 *Let*

(H_{30}) $\displaystyle\min_{0 \le t \le T} f^\infty = \infty$

and

(H_{31}) $\displaystyle\max_{0 \le t \le T} f^0 = 0.$

Then Eq. (4.1) has at least two positive T-periodic solutions.

Proof From (H_{30}), it follows that there exists $c_2 > 0$ large enough such that $f(t, x) \ge a(t)x$ for $c_2 \le x \le \frac{c_2}{\delta}$. Define ψ as in the proof of Theorem 4.1.1 and set

$c_3 = \frac{c_2}{\delta}$ and $\phi_0(t) = \phi_0 = \frac{c_2+c_3}{2}$. Then $\phi_0 \in \{x \in K(\psi, c_2, c_3) : \psi(x) > c_2\}$. For $x \in K(\psi, c_2, c_3)$, we have

$$\psi(Ax) = \min_{0 \le t \le T} \int_t^{t+T} G(t, s) f(s, x(h(s))) \, ds$$

$$\ge \min_{0 \le t \le T} \int_t^{t+T} G(t, s) a(s) x(s) \, ds$$

$$\ge c_2 \min_{0 \le t \le T} \int_t^{t+T} a(s) G(t, s) \, ds$$

$$= c_2.$$

Next, by (H_{31}), there exists ξ, $0 < \xi < c_2$ such that $f(t, x) < a(t)x$ for $0 < x < \xi$. Set $c_1 = \xi$. Then $c_1 < c_2$ and $f(t, x) < a(t)c_1$ for $0 < x < c_1$. Now, for $x \in \overline{K}_{c_1}$, we have

$$\|Ax\| = \sup_{0 \le t \le T} \int_t^{t+T} G(t, s) f(s, x(h(s))) \, ds$$

$$\le \sup_{0 \le t \le T} \int_t^{t+T} G(t, s) a(s) x(s) \, ds$$

$$< c_1 \sup_{0 \le t \le T} \int_t^{t+T} a(s) G(t, s) \, ds$$

$$= c_1.$$

In addition, for $x \in \overline{K}_{c_3}$ with $\|Ax\| > c_3$, we have

$$\psi(Ax) = \min_{0 \le t \le T} \int_t^{t+T} G(t, s) f(s, x(h(s))) \, ds$$

$$\ge \alpha \int_0^T f(s, x(h(s))) \, ds$$

and

$$c_3 < \|Ax\| \leq \beta \int_0^T f(s, x(h(s))) \, ds$$

$$= \frac{\alpha}{\delta} \int_0^T f(s, x(h(s))) \, ds$$

$$\leq \frac{1}{\delta} \psi(Ax).$$

The above inequalities imply that

$$\psi(Ax) > \frac{c_2}{c_3} \|Ax\|.$$

Hence, by Theorem 1.2.1, Eq. (4.1) has at least two positive T-periodic solutions. This completes the proof of the theorem.   □

Theorem 4.1.4 *Suppose that there exists a constant μ, $0 < \mu \leq 1$ such that*

(H_{32}) $f^\infty > \dfrac{1}{\mu}$

and

(H_{33}) $f^0 < \mu$.

Then there exist at least two positive T-periodic solutions of Eq. (4.1).

Proof Since (H_{32}) holds, there exists $c_2 > 0$ such that

$$f(t, x) > \frac{a(t)x}{\mu} \quad \text{for} \quad c_2 \leq x \leq \frac{c_2}{\delta}.$$

Define the nonnegative concave continuous functional ψ on K by $\psi(x) = \min_{t \in [0,T]} x(t)$. Take $c_3 = \frac{c_2}{\delta}$ and $\phi_0(t) = \frac{c_2 + c_3}{2}$. This shows that $\phi_0(t) \in \{x : x \in K(\psi, c_2, c_3), \psi(x) > c_2\} \neq \emptyset$. Then for $x \in K(\psi, c_2, c_3)$, we have

$$\psi(Ax) = \min_{0 \leq t \leq T} \int_t^{t+T} G(t, s) f(s, x(h(s))) \, ds$$

$$> \min_{0 \leq t \leq T} \int_t^{t+T} G(t, s) \frac{a(s)x(s)}{\mu} \, ds$$

$$\geq \frac{c_2}{\mu} \min_{0 \leq t \leq T} \int_t^{t+T} a(s)G(t, s) \, ds$$

$$> c_2.$$

From (H_{33}), there exists a real ξ, $0 < \xi < c_2$ such that $f(t, x) < a(t)\mu x$ for $0 < x \leq \xi$. Set $c_1 = \xi$; then $c_1 < c_2$. For $x \in \overline{K}_{c_1}$, we have

$$\|Ax\| = \sup_{0 \leq t \leq T} \int_t^{t+T} G(t, s)f(s, x(h(s))) \, ds$$

$$< \sup_{0 \leq t \leq T} \int_t^{t+T} G(t, s)a(s)\mu\|x\| \, ds$$

$$\leq \mu c_1$$

$$< c_1.$$

The rest of the proof is similar to that of Theorem 4.1.1 and is omitted.   □

Corollary 4.1.1 *If $f^0 < 1$ and $f^\infty > 1$, then Eq. (4.1) has at least two positive T-periodic solutions.*

Theorem 4.1.5 *If*

(H_{34}) $\displaystyle \max_{t \in [0,T]} F^0 = \alpha_1 \in \left(0, \frac{1}{\beta T}\right)$

and there exists a constant $c_2 > 0$ such that

(H_{35}) $f(t, x) > \dfrac{1}{\alpha \delta T} x$ *for* $c_2 \leq x \leq \dfrac{c_2}{\delta}$,

then Eq. (4.1) has at least two positive T-periodic solutions.

Remark 4.1.1 The conditions in Theorems 4.1.1–4.1.4, Corollary 4.1.1, and Theorem 4.1.5 improve the results in [4, 8, 15, 16].

Theorem 4.1.6 *Suppose that*
(H_{36}) *f is nondecreasing with respect to x*

and there are constants $0 < c_1 < c_2$ such that

(H_{37}) $\dfrac{\int_0^T f(t, c_1) \, dt}{(1 - \delta)c_1} < 1 < \dfrac{\delta \int_0^T f(t, \delta c_2) \, dt}{(1 - \delta)c_2}.$

Then Eq. (4.1) has at least two positive T-periodic solutions.

Proof Set $c_3 = \frac{c_2}{\delta}$, define ψ as in the proof of Theorem 4.1.1, and let $\phi_0(t) = \phi_0 = \frac{c_2 + c_3}{2}$. Then $\phi_0 \in \{x \in K(\psi, c_2, c_3) : \psi(x) > c_2\}$. For $x \in K(\psi, c_2, c_3)$, applying (H_{36}) and (H_{37}), we obtain

$$
\begin{aligned}
\psi(Ax) &= \min_{0 \le t \le T} \int_t^{t+T} G(t, s) f(s, x(h(s))) \, ds \\
&\ge \frac{\delta}{1 - \delta} \int_0^T f(s, x(h(s))) \, ds \\
&\ge \frac{\delta}{1 - \delta} \int_0^T f(s, \delta c_2) \, ds \\
&> c_2.
\end{aligned}
$$

Next, for $x \in \overline{K}_{c_1}$, we have

$$
\begin{aligned}
\|Ax\| &= \sup_{0 \le t \le T} \int_t^{t+T} G(t, s) f(s, x(h(s))) \, ds \\
&\le \frac{1}{1 - \delta} \int_0^T f(s, \|x\|) \, ds \\
&\le \frac{1}{1 - \delta} \int_0^T f(s, c_1) \, ds \\
&< c_1
\end{aligned}
$$

by using (H_{36}) and (H_{37}). Finally, for $x \in \overline{K}_{c_3}$ with $\|Ax\| > c_3$, we have

$$
\begin{aligned}
\psi(Ax) &= \min_{0 \le t \le T} \int_t^{t+T} G(t, s) f(s, x(h(s))) \, ds \\
&\ge \frac{\delta}{1 - \delta} \int_0^T f(s, x(h(s))) \, ds
\end{aligned}
$$

and

$$c_3 < \|Ax\| \le \frac{1}{1-\delta} \int_0^T f(s, x(h(s)))\, ds$$

$$\le \frac{1}{\delta} \psi(Ax),$$

which together imply

$$\psi(Ax) > \frac{c_2}{c_3} \|Ax\|.$$

Thus, all the conditions of Theorem 1.2.1 are satisfied and so Eq. (4.1) has at least two positive T-periodic solutions. This completes the proof of the theorem.   □

Theorem 4.1.7 *Suppose that* (H_{36}) *holds and there are constants* $0 < c_1 < c_2$ *such that*

$$(H_{38}) \quad \frac{T \max\limits_{t \in [0,T]} f(t, c_1)}{(1-\delta)c_1} < 1 < \frac{\delta\, T \min\limits_{t \in [0,T]} f(t, \delta c_2)}{(1-\delta)c_2}.$$

Then Eq. (4.1) has at least two positive T-periodic solutions.

Proof Take ψ as in the proof of Theorem 4.1.1 and let $\phi_0(t) = \frac{c_2+c_3}{2}$, where $c_3 = \frac{c_2}{\delta}$. Then $\phi_0(t) \in \{x \in K(\psi, c_2, c_3) : \psi(x) > c_2\} \ne \phi$. Now using (H_{36}) and (H_{38}), we have for $x \in K(\psi, c_2, c_3)$,

$$\psi(Ax) = \min\limits_{0 \le t \le T} \int_t^{t+T} G(t,s) f(s, x(h(s)))\, ds$$

$$\ge \frac{\delta}{1-\delta} \int_0^T f(s, \delta c_2)\, ds$$

$$\ge \frac{\delta}{1-\delta} \min\limits_{0 \le t \le T} f(t, \delta c_2)\, T$$

$$> c_2.$$

For $x \in \overline{K}_{c_1}$, we can use (H_{36}) and (H_{38}) to obtain

$$\|Ax\| = \sup\limits_{0 \le t \le T} \int_t^{t+T} G(t,s) f(s, x(h(s)))\, ds$$

$$\le \frac{1}{1-\delta} \int_0^T f(s, \|x\|)\, ds$$

$$\leq \frac{1}{1 - \delta} \max_{0 \leq t \leq T} f(t, c_1) T$$

$$< c_1.$$

The last part of the proof is similar to that of Theorem 4.1.1 and hence is omitted. Therefore, (4.1) has at least two positive T-periodic solutions and this completes the proof of the theorem.   □

Wang [12] considered the differential equation

$$x'(t) = a(t)g(x(t))x(t) - \lambda b(t) f(x(t - \tau(t))), \tag{4.9}$$

where $\lambda > 0$ is a positive parameter, $a, b \in C(R, [0, \infty))$ are T-periodic functions, $\int_0^T a(t)\, dt > 0$, $\int_0^T b(t)\, dt > 0$, $\tau \in C(R, R)$ is a T-periodic function, $f, g : [0, \infty) \to [0, \infty)$ are continuous, $0 < l \leq g(x) < L < \infty$ for $x \geq 0$, l, L are positive constants and $f(x) > 0$ for $x > 0$. In developing sufficient conditions for the existence of positive T-periodic solutions he introduced the notations

$$i_0 = \textit{number of zeros in the set } \{\overline{f}_0, \overline{f}_\infty\}$$

and

$$i_\infty = \textit{number of infinities in the set } \{\overline{f}_0, \overline{f}_\infty\},$$

where

$$\overline{f}_0 = \lim_{x \to 0^+} \frac{f(x)}{x} \quad \textit{and} \quad \overline{f}_\infty = \lim_{x \to \infty} \frac{f(x)}{x}.$$

In what follows, we apply Theorem 1.2.2 to Eq. (4.9) to obtain some new results different from those in [12]. The Banach space X and a cone K are same as defined earlier in the chapter but the operator A is replaced by

$$(A_\lambda x)(t) = \lambda \int_t^{t+T} G_x(t, s) b(s) f(x(s - \tau(s)))\, ds,$$

where

$$G_x(t, s) = \frac{e^{-\int_t^s a(\theta) g(x(\theta))\, d\theta}}{1 - e^{-\int_0^T a(\theta) g(x(\theta))\, d\theta}}$$

is the Green's kernel. The Green's kernel $G_x(t, s)$ satisfies the property

$$\frac{\delta^L}{1 - \delta^L} \leq G_x(t, s) \leq \frac{1}{1 - \delta^l}.$$

Proceeding as in the proof of Theorem 4.1.6, we obtain the following result.

Theorem 4.1.8 *Let (H_{36}) hold. Further, assume that there are constants $0 < c_1 < c_2$ such that*

$$(H_{39}) \quad \frac{(1 - \delta^L)c_2}{\delta^L f(c_2) \int\limits_0^T b(s)\,ds} < \lambda < \frac{(1 - \delta^l)c_1}{f(c_1) \int_0^T b(s)\,ds}.$$

Then Eq. (4.9) has at least two positive T-periodic solutions.

Section 2.2 of Chap. 2 deals with the existence of at least three positive T-periodic solutions of the Eq. (4.1) with a parameter λ. Some of the results can be extended to Eq. (4.9). In the following, we apply Theorem 1.2.2 to Eq. (4.9) to obtain a different sufficient condition for the existence of at least three positive T-periodic solutions.

Theorem 4.1.9 *Let $\overline{f}_0 < 1 - \delta^l$ and $\overline{f}_\infty < 1 - \delta^l$ hold. Assume that there exists a constant $c_2 > 0$ such that*

$$(H_{40}) \quad f(x) > \frac{(1 - \delta^L)}{\delta^{2L}}c_2 \ for \ c_2 \le x \le \frac{(1 - \delta^L)}{\delta^L(1 - \delta^l)}c_2.$$

Then Eq. (4.9) has at least three positive T-periodic solutions for

$$\frac{\delta^L}{\int\limits_0^T b(t)\,dt} < \lambda < \frac{1}{\int\limits_0^T b(t)\,dt}.$$

Proof Since $\bar{f}^\infty < 1 - \delta^l$, there exist $0 < \epsilon < 1 - \delta^l$ and $\xi > 0$ such that $f(x) \le \epsilon x$ for $x \ge \xi$. Let $\gamma = \max\limits_{0 \le x \le \xi, 0 \le t \le T} f(x)$. Then $f(x) \le \epsilon x + \gamma$ for $x \ge 0$.

Choose $c_4 > 0$ such that

$$c_4 > \max\left\{ \frac{\gamma}{(1 - \delta^l) - \epsilon}, \frac{1 - \delta^L}{\delta^L(1 - \delta^l)}c_2 \right\}.$$

Then, for $x \in \overline{K}_{c_4}$,

$$\|A_\lambda x\| = \sup_{0 \le t \le T} \lambda \int\limits_t^{t+T} G(t, s)b(s)f(x(s - \tau(s)))\,ds$$

$$\le \frac{1}{1 - \delta^l} \lambda \int\limits_0^T b(s)f(x(s - \tau(s)))\,ds$$

$$\leq \frac{1}{1 - \delta^l} \lambda \int_0^T b(s)(\epsilon \|x\| + \gamma) \, ds$$

$$\leq \frac{1}{1 - \delta^l}(\epsilon c_4 + \gamma)$$

$$< c_4,$$

that is, $A : \overline{K}_{c_4} \to \overline{K}_{c_4}$.

Now, we define a nonnegative concave continuous functional ψ on K by $\psi(x) = \min_{t \in [0,T]} x(t)$. Then $\psi(x) \leq \|x\|$. Set $c_3 = \frac{1 - \delta^L}{\delta^L(1 - \delta^l)} c_2$ and $\phi_0(t) = \phi_0 = \frac{c_2 + c_3}{2}$. Then $c_2 < c_3$ and $\phi_0 \in \{x \in K(\psi, c_2, c_3) : \psi(x) > c_2\}$. For $x \in K(\psi, c_2, c_3)$, it follows from (H_{40}) that

$$\psi(A_\lambda x) = \min_{0 \leq t \leq T} \lambda \int_t^{t+T} G(t, s)b(s)f(x(s - \tau(s))) \, ds$$

$$\geq \frac{\delta^L}{1 - \delta^L} \lambda \int_0^T b(s)f(x(s - \tau(s))) \, ds$$

$$\geq \frac{\delta^L}{1 - \delta^L} \lambda \int_0^T b(s) \frac{1 - \delta^L}{\delta^{2L}} c_2 \, ds$$

$$> c_2.$$

Next, since $\bar{f}^0 < 1 - \delta^l$, there exists a positive $\sigma < c_2$ such that

$$f(x) < (1 - \delta^l)x \quad \text{for } 0 < x \leq \sigma.$$

Set $c_1 = \sigma$; then $c_1 < c_2$. For $x \in \overline{K}_{c_1}$, we have

$$\|A_\lambda x\| = \sup_{0 \leq t \leq T} \lambda \int_t^{t+T} G(t, s)b(s)f(x(s - \tau(s))) \, ds$$

$$\leq \frac{1}{1 - \delta^l} \lambda \int_0^T b(s)(1 - \delta^l)\|x\| \, ds$$

$$\leq \frac{1}{1 - \delta^l} \lambda \int_0^T b(s)(1 - \delta^l)c_1 \, ds$$

$$< c_1.$$

Finally, for $x \in K(\psi, c_2, c_4)$ with $\|A_\lambda x\| > c_3$, we have

$$c_3 < \|A_\lambda x\| \leq \frac{1}{1 - \delta^l} \lambda \int_0^T b(s) f(x(s - \tau(s))) \, ds,$$

which, in turn implies that

$$\psi(A_\lambda x) \geq \frac{\delta^L}{1 - \delta^L} \lambda \int_0^T b(s) f(x(s - \tau(s))) \, ds$$

$$> \frac{\delta^L}{1 - \delta^L} (1 - \delta^l) c_3$$

$$= c_2.$$

Hence, by Theorem 1.2.2, Eq. (4.9) has at least three positive T-periodic solutions. $\square$

Corollary 4.1.2 *If $i_0 = 2$ and there exists a constant $c_2 > 0$ such that (H_{40}) holds, then Eq. (4.9) has at least three positive T-periodic solutions.*

Remark 4.1.2 Wang [12] obtained three different results for the existence of at least one positive periodic solution of (4.9) using fixed point index theory [2]. In Corollary 4.1.2, it has been shown that (4.9) has at least three positive T-periodic solutions when $i_0 = 2$.

It would be interesting to obtain sufficient conditions for the existence of at least two or three positive periodic solutions of (4.9) when $i_0 \in \{0, 1\}$ and $i_0 \in \{0, 1, 2\}$ by using the Leggett-Williams multiple fixed point theorems. Bai and Xu [1] obtained a sufficient condition ([1, Theorem 3.2]) for the existence of three nonnegative T-periodic solutions of (4.9). Although the condition $i_0 = 2$ holds both in [1, Theorem 3.2] and in Corollary 4.1.2 above, condition (H_{40}) and the condition (H_5) in [1] are different. Accordingly, the ranges on the parameter λ are also different.

Finally, we generalize some of the above results to the scalar differential equation of the form

$$\frac{dx}{dt} = -A(t)x(t) + f(t, x(t)), \tag{4.10}$$

where $A \in C(R, R)$ and $f \in C(R \times R, R)$ satisfy $A(t + T) = A(t)$ and $f(t + T, x) = f(t, x)$. We shall apply Theorem 1.2.1 to obtain the existence of at least two positive periodic solutions of (4.10).

Lemma 4.1.1 *If $x(t)$ is a T—periodic solution of (4.10) then it satisfies the integral equation*

$$x(t) = \int_t^{t+T} G(t, s) f(s, x(s)) \, ds \tag{4.11}$$

where $G(t, s)$ is the Green's function given by

$$G(t, s) = \frac{\exp\left(\int\limits_t^s A(\theta)d\theta\right)}{\exp\left(\int\limits_0^T A(\theta)d\theta\right) - 1}, \quad t, s \in R. \tag{4.12}$$

Now, let us define

$$\delta = \exp\left(\int\limits_0^T A(\theta)d\theta\right). \tag{4.13}$$

Observe that $\delta > 1$ if

$$\int\limits_0^T A(\theta)d\theta > 0. \tag{4.14}$$

Under the assumption (4.14), the Green's function (4.12) satisfies

$$0 < \frac{1}{\delta - 1} < G(t, s) < \frac{\delta}{\delta - 1}, \quad s \in [t, t + T]. \tag{4.15}$$

We know that the set

$$X = \{x \in C([0, T], R) : x(0) = x(T)\} \tag{4.16}$$

endowed with the norm

$$\|x\| = \sup_{0 \le t \le T} x(t) \tag{4.17}$$

is a Banach space where $C[0, T]$ is the set of all continuous functions defined on $[0, T]$.

Theorem 4.1.10 *Let $\int_0^T A(s)ds > 0$. Assume:*

(H_{41}) *there exists $c_3 > 0$ such that $\int_0^T f(s, x)ds > 0$ if $0 < x \le c_3$ and*

$$\int\limits_0^T f(s, x)ds \ge \frac{\delta - 1}{\delta}c_3 \quad \text{if } \frac{c_3}{\delta} \le x \le c_3; \tag{4.18}$$

(H_{42}) $\lim\limits_{\|x\| \to 0} \frac{1}{\|x\|} \int\limits_0^T f(s, x)ds < \frac{\delta - 1}{\delta}.$

Then Eq. (4.10) has at least two positive T-periodic solutions in $\overline{K}_{c_3}$.

Proof Let us consider the Banach space X endowed with the sup norm as defined in (4.16)–(4.17). Define a cone K on X by

$$K = \{x \in X : x(t) > 0\}. \tag{4.19}$$

Let c_3 be a positive constant satisfying the conditions in the hypotheses. Define an operator $E : \overline{K}_{c_3} \to K$ by

$$(Ex)(t) = \int_t^{t+T} G(t, s) f(s, x(s)) \, ds. \tag{4.20}$$

It is clear that the existence of a fixed point of E is equivalent to the existence of a positive periodic solution of (4.10).

We shall apply Leggett-Williams multiple fixed point theorem to the above operator E to prove the existence of at least two positive periodic solutions for the Eq. (4.10).

It can be easily verified that E is well defined, completely continuous on $\overline{K}_{c_3}$, and $E(\overline{K}_{c_3}) \subset K$. Consider the nonnegative concave continuous functional ψ defined on K by

$$\psi(x) = \min_{0 \le t \le T} x(t). \tag{4.21}$$

For $c_2 = \frac{c_3}{\delta}$ and $\phi_0 = \frac{1}{2}(c_2 + c_3)$ we have, $c_2 < \phi_0 < c_3$ and so

$$\{x \in K(\psi, c_2, c_3) : \psi(x) > c_2\} \ne \emptyset.$$

For $x(t) \in K(\psi, c_2, c_3)$,

$$\psi(Ex) = \min_{0 \le t \le T} \int_t^{t+T} G(t, s) f(s, x(s)) \, ds$$

$$> \frac{1}{\delta - 1} \int_0^T f(s, x(s)) \, ds \quad \text{(from (4.15))}$$

$$\ge \frac{1}{\delta - 1} \frac{\delta - 1}{\delta} c_3 \quad \text{(from H}_{41})$$

$$= \frac{c_3}{\delta}.$$

Hence, condition (i) of Theorem 1.2.1 is satisfied.

Now, we show that condition (ii) of Theorem 1.2.1 holds. From condition (H_{42}), there exists ξ, $0 < \xi < c_2$, such that

$$\int_0^T f(s, x(s))\, ds < \frac{\delta - 1}{\delta} \|x\| \quad \text{for } 0 \le \|x\| \le \xi.$$ (4.22)

Choose $c_1 = \xi$. Then we have $0 < c_1 < c_2$ and for $0 \le x(s) \le c_1$, applying (4.15) and (4.22), we obtain

$$\begin{aligned}
\|Ex\| &= \sup_{0 \le t \le T} \int_0^T G(t, s) f(s, x(s))\, ds \\
&< \frac{\delta}{\delta - 1} \int_0^T f(s, x(s))\, ds \\
&\le \frac{\delta}{\delta - 1} \frac{\delta - 1}{\delta} \|x\| \\
&\le c_1.
\end{aligned}$$

Hence, condition (ii) in Theorem 1.2.1 is established.
  Now from (4.15),

$$\begin{aligned}
\psi(Ex) &= \min_{0 \le t \le T} \int_t^{t+T} G(t, s) f(s, x(s))\, ds \\
&> \frac{1}{\delta - 1} \int_0^T f(s, x(s))\, ds.
\end{aligned}$$ (4.23)

Let $0 < x(t) \le c_3$ be such that $\|Ex\| > c_3$. For such a choice of $x(t)$, we have

$$\begin{aligned}
c_3 < \|Ex\| &= \sup_{0 \le t \le T} \int_0^T G(t, s) f(s, x(s))\, ds \\
&< \frac{\delta}{\delta - 1} \int_0^T f(s, x(s))\, ds \\
&\le \delta \frac{1}{\delta - 1} \int_0^T f(s, x(s))\, ds \\
&< \delta \psi(Ex)
\end{aligned}$$

by (4.23). Therefore, $\psi(Ex) > \frac{1}{\delta} \|Ex\|$ and this implies that $\psi(Ex) > \frac{c_2}{c_3} \|Ex\|$ for $0 < x(t) \le c_3$ satisfying $\|Ex\| > c_3$. Hence, condition (iii) of Theorem 1.2.1 is also satisfied. Therefore, by Theorem 1.2.1, the operator (4.20) has at least two fixed points in $\overline{K}_{c_3}$, so Eq. (4.10) admits at least two positive T-periodic solutions. This completes the proof. □

Corollary 4.1.3 *Let $\int\limits_0^T A(s)\,ds > 0$. Assume there exists a positive constant c_3 such that*

$$\int\limits_0^T f(t, x)dt > 0 \ for \ 0 < x \leq c_3, \tag{4.24}$$

$$\int\limits_0^T f(s, x)ds = \frac{\delta - 1}{\delta} x \ for \ x = c_3, \tag{4.25}$$

$$\int\limits_0^T f(s, x)ds > \frac{\delta - 1}{\delta} c_3 \ for \ \frac{c_3}{\delta} \leq x < c_3, \tag{4.26}$$

and

$$(H_{42}^*) \qquad \lim_{x \to 0} \frac{1}{x} \int\limits_0^T f(s, x)ds < \frac{\delta-1}{\delta}.$$

Then Eq. (4.10) has at least two positive T-periodic solutions in $\overline{K}_{c_3}$.

Proof Assume that there exists $c_3 > 0$ such that (4.24)–(4.26) hold. This implies that

$$\int\limits_0^T f(s, x)\,ds \geq \frac{\delta - 1}{\delta} c_3 \ for \ \frac{c_3}{\delta} \leq x \leq c_3,$$

and hence (H_{42}^*) implies (H_{42}).

Now, let us assume

$$\lim_{x \to 0} \int\limits_0^T \frac{f(s, x)}{x}\,ds < \frac{\delta - 1}{\delta}. \tag{4.27}$$

We have $\frac{1}{\|x\|} \int_0^T f(s, x)\,ds = \int_0^T \frac{f(s,x)}{\|x\|}\,ds \leq \int_0^T \frac{f(s,x)}{x(s)}\,ds$ for $s \in [0, T]$. Observe that $\|x\| \to 0$ if and only if $x(s)$ also tends to zero for all $s \in [0, T]$. Therefore, in view of (4.27) we have

$$\lim_{\|x\| \to 0} \frac{1}{\|x\|} \int_0^T f(s, x(s)) \, ds \leq \lim_{x(s) \to 0} \int_0^T \frac{f(s, x(s))}{x(s)} \, ds < \frac{\delta - 1}{\delta}, \quad \text{for all } s \in [0, T].$$

Therefore, condition (H_{42}^*) implies (H_{42}), and this completes the proof of the theorem. □

4.2 Applications to Some Mathematical Models

Ye et al. [15] and Zhang et al. [16] showed that the models (4.3)–(4.6) have at least one positive periodic solution. In the following section, we apply some of the results obtained in Sect. 4.1 to obtain sufficient conditions for the existence of at least two positive periodic solutions of the models (4.3)–(4.6).

Example 4.2.1 The generalized logistic model for a single species

$$x'(t) = x(t)[a(t) - b(t)x(t) - c(t)x(t - \tau(t))] \tag{4.28}$$

has at least two positive T-periodic solutions, where $a(t)$, $b(t)$ and $c(t)$ are nonnegative continuous periodic functions.

To see this, set $f(t, x) = x(t)[b(t)x(t) + c(t)x(t - \tau(t))]$. Since

$$\max_{t \in [0,T]} \frac{f(t, x)}{a(t)x} \leq \max_{t \in [0,T]} \left\{ \frac{b(t)}{a(t)} \right\} \|x\| + \max_{t \in [0,T]} \left\{ \frac{c(t)}{a(t)} \right\} \|x\| \to 0 \text{ as } x \to 0,$$

we see that (H_{31}) is satisfied. Moreover,

$$\min_{t \in [0,T]} \frac{f(t, x)}{a(t)x} \geq \min_{t \in [0,T]} \delta \left\{ \frac{b(t)}{a(t)} \right\} \|x\| \to \infty \text{ as } x \to \infty;$$

so (H_{30}) is satisfied. Thus, by Theorem 4.1.3, Eq. (4.28) has at least two positive T-periodic solutions.

Example 4.2.2 The logistic equation for a single species

$$x'(t) = x(t) \left[a(t) - \sum_{i=1}^n b_i(t)x(t - \tau_i(t)) \right] \tag{4.29}$$

has at least two positive T-periodic solutions, where a, b_i, $\tau_i \in C(R, R_+)$ are T-periodic functions.

Example 4.2.3 The logistic equation with several delays (4.3) has at least two positive T-periodic solutions.

Example 4.2.4 The generalized Richards single species growth model (4.4) has at least two positive T-periodic solutions.

The verification of Examples 4.2.2–4.4.4 are similar to that of Example 4.2.1.

Applying Corollary 4.1.1 to the generalized Michaelis-Menton type single species growth model (4.5), we obtain the following result.

Example 4.2.5 If

$$\min_{t\in[0,T]} \sum_{i=1}^{n} \frac{b_i(t)}{a(t)c_i(t)} > 1,$$

then (4.5) has at least two positive T-periodic solutions.

Now, we assume that the population is subject to harvesting. Under the catch-per-unit-effort hypothesis [15], the harvested population's growth model becomes

$$x'(t) = x(t)\left[a(t) - \frac{b(t)x(t)}{1 + c(t)x(t)}\right] - qEx, \qquad (4.30)$$

where q and E are positive constants denoting the catch-ability coefficients and harvesting effort, respectively. Ye et al. [15] proved that if $0 < qE < \frac{1-\delta}{T}$ and $\left(\frac{b^m}{c} + qE\right) > \frac{1-\delta}{\delta^2 T}$, then (4.30) has at least one positive T-periodic solution, where $b^m = \min_{0 \le t \le T} b(t)$ and $0 < c(t) \le c$.

Theorem 4.2.1 *Suppose that* $0 < qE < \frac{1-\delta}{T}$ *and* $\dfrac{\delta \int_0^T b(t)\,dt}{c} + qE > \frac{1-\delta}{\delta^2 T}$. *Then* Eq.(4.30) *has at least two positive T-periodic solutions.*

Proof Set $f(t,x) = \frac{b(t)x^2}{1+c(t)x} + qEx$. Then $qE < \frac{1-\delta}{T}$ implies the condition (H_{34}).

Choose $c_2 = \dfrac{\delta(1-qE\alpha\delta T)}{\alpha\delta^2 T \int_0^T b(t)\,dt - c(1-qE\alpha\delta T)}$; then $\dfrac{c_2\alpha\delta^2 T \int_0^T b(t)\,dt}{\delta + cc_2} + qE\alpha\delta T = 1$. Setting

$c_3 = \frac{c_2}{\delta} = \dfrac{1-qE\alpha\delta T}{\alpha\delta^2 T \int_0^T b(t)\,dt - c(1-qE\alpha\delta T)}$, we have $c_2 < c_3$. Now for $c_2 \le x \le \frac{c_3}{\delta}$, we have

$$f(t,x) > \dfrac{c_2^2 \int_0^T b(t)\,dt}{1 + c\frac{c_2}{\delta}} + qEc_2$$

$$= \dfrac{c_2}{\alpha\delta T}\left[\dfrac{c_2\alpha\delta^2 T \int_0^T b(t)\,dt}{\delta + cc_2} + qE\alpha\delta T\right]$$

$$\ge \dfrac{c_2}{\alpha\delta T},$$

that is, (H_{35}) holds. Hence, by Theorem 4.1.5, (4.30) has at least two positive T-periodic solutions. □

Remark 4.2.1 There are very few results in the literature on the existence of two periodic solutions of (4.1) with its application to the models (4.3)–(4.6). Hence, simple results on the existence of two periodic solutions of the above equations are of immense importance.

4.3 Application to Renewable Resource Dynamics

In this section, we apply Theorem 4.1.10 and Corollary 4.1.3 to investigate the existence of positive T-periodic solutions of the ordinary differential equation

$$x' = a(t)x(x - b(t))(c(t) - x) \tag{4.31}$$

representing dynamics of a renewable resource that is subjected to Allee effects. The transformation $y(t) = c(t)x(t)$ transforms Eq. (4.31) in to

$$\frac{dy}{dt} = -\left(a(t)c^2(t)k(t) + \frac{c'(t)}{c(t)}\right)y + a(t)c^2(t)\left((1 + k(t)) - y\right)y^2 \tag{4.32}$$

where

$$k(t) = \frac{b(t)}{c(t)} < 1. \tag{4.33}$$

Note that (4.32) is a particular case of a general scalar differential equation of the form

$$\frac{dy}{dt} = -A(t)y(t) + f(t, y(t)) \tag{4.34}$$

where $A \in C(R, R)$ and $f \in C(R \times R, R)$ satisfy $A(t+T) = A(t)$ and $f(t+T, x) = f(t, x)$. Comparing (4.32) with (4.34) we have

$$A(t) = \left(a(t)c^2(t)k(t) + \frac{c'(t)}{c(t)}\right) \tag{4.35}$$

and

$$f(t, x) = a(t)c^2(t)\left((1 + k(t)) - x\right)x^2. \tag{4.36}$$

Let us consider a Banach space X as defined in (4.16)–(4.17). We have $f(t, 0) = 0$, $f(t, x(t)) > 0$ for $0 < x(t) < 1 + k_m$, and $f(t, x(t)) < 0$ for $x(t) > 1 + k_M$ where $k_m = \min_{0 \le t \le T} k(t)$ and $k_M = \max_{0 \le t \le T} k(t)$.

Henceforth, we let

$$M = \int_0^T a(s)c^2(s)\,ds \quad \text{and} \quad N = \int_0^T a(s)c^2(s)k(s)\,ds. \tag{4.37}$$

Since $0 < k(t) < 1$ (from 4.33), we have $M > N > 0$. From (4.36) we observe that $\lim_{x \to 0} \frac{1}{x} \int_0^T f(s, x)\,ds = 0$ and hence (H_{42}^*) of Corollary 4.1.3 is satisfied for Eq. (4.32). We then have the following theorem.

Theorem 4.3.1 *If*

$$\frac{(M + N) + \sqrt{(M + N)^2 - 4M(\frac{e^N - 1}{e^N})}}{2M} > \frac{e^{2N} - \frac{1}{e^N}}{M + N} \tag{4.38}$$

then Eq. (4.31) has at least two positive T-periodic solutions.

Proof We shall use Corollary 4.1.3 to prove this theorem. From (4.35), it is easy to observe that $\int_0^T A(s)\,ds = N > 0$. To complete the proof of the theorem, it suffices to find the existence of a positive constant $c_3 > 0$ such that (4.24)–(4.26) hold.

Take

$$c_3 = \frac{(M + N) + \sqrt{(M + N)^2 - 4M(\frac{\delta - 1}{\delta})}}{2M} \tag{4.39}$$

and define $c_2 = \frac{c_3}{\delta}$. Clearly $0 < c_2 < c_3$. It is easy to verify that $p = c_3$ is a solution of

$$-Mp^2 + (M + N)p - \frac{\delta - 1}{\delta} = 0. \tag{4.40}$$

A simple calculation shows that (4.40) is equivalent to

$$(1 - p)p \int_0^T a(s)c^2(s)\,ds + p \int_0^T a(s)c^2(s)k(s)\,ds = \frac{\delta - 1}{\delta}. \tag{4.41}$$

The above equation can be rewritten as

$$\int_0^T f(s, p)\,ds = \frac{\delta - 1}{\delta}p. \tag{4.42}$$

That is, $p = c_3$ satisfies

$$\int_0^T f(s, c_3)\, ds = \frac{\delta - 1}{\delta} c_3.$$

Next, we consider the inequality

$$\int_0^T f\left(s, \frac{c_3}{\delta}\right) ds > \frac{\delta - 1}{\delta} c_3. \qquad (4.43)$$

Substituting for f, we obtain

$$\int_0^T a(s) c^2(s)\left(1 + k(s) - \frac{c_3}{\delta}\right)\frac{c_3^2}{\delta^2}\, ds > \frac{\delta - 1}{\delta} c_3. \qquad (4.44)$$

The above inequality is equivalent to

$$-M c_3^2 + (M + N)\delta c_3 - \delta^2(\delta - 1) > 0.$$

Since $p = c_3$ is a solution of (4.40), the above inequality yields

$$c_3 > \frac{\delta^2 - \frac{1}{\delta}}{(M + N)}. \qquad (4.45)$$

Therefore, (4.43) will be satisfied if the root $p = c_3$ of (4.40) satisfies the inequality (4.45). Thus, (4.24)–(4.26) will be satisfied if the parameters of the associated Eq. (4.32) satisfies (4.45), that is, (4.38) holds. Hence, the proof is complete. □

Observe that Theorem 4.3.1 verifiable only if M and N satisfy the inequality

$$(M + N)^2 - 4M\left(\frac{e^N - 1}{e^N}\right) > 0. \qquad (4.46)$$

Note that the left hand side of the inequality (4.46) is an implicit expression in M and N and the Fig. 4.1 presents the region in the (M, N) space where the inequality is satisfied. From this figure we observe that (4.46) is valid in the interior of the positive quadrant of (M, N) space. Since we have both M and N to be positive from (4.37), we see that (4.46) is always satisfied for the model (4.31).

Figure 4.2 presents the region in the (M, N) space where the inequality (4.38) is satisfied. This figure helps in identifying the coefficient functions that ensure the existence of at least two T-periodic solutions for (4.31). Let us choose the functions $a(t)$, $b(t)$ and $c(t)$ to be the 2π periodic functions

Fig. 4.1 The shaded region represents the portion in the (M, N) space that satisfies (4.46)

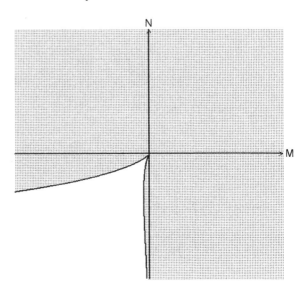

$$a(t) = (1.2 + \sin t)^2, \quad b(t) = \frac{1.2 + \cos t}{12(1.2 + \sin t)}, \quad \text{and} \quad c(t) = \frac{1}{1.2 + \sin t}. \quad (4.47)$$

From (4.33), we have $k(t) = \frac{b(t)}{c(t)} = \frac{1.2 + \cos t}{12} < 1$. According to (4.37), we have $M = 6.28, \ N = 0.628$. Clearly, $M > N > 0$. Also, we have

$$\frac{(M + N) + \sqrt{(M + N)^2 - 4M \left(\frac{e^N - 1}{e^N} \right)}}{2M} = 1.0277$$

and

$$\frac{e^{2N} - \frac{1}{e^N}}{M + N} = 0.4310.$$

Therefore (4.38) is satisfied and hence (4.31) admits at least two positive solutions with $a(t)$, $b(t)$, and $c(t)$ as given in (4.47). The existence of 2π periodic solutions can also be ascertained from Fig. 4.2 by observing presence of the point $(M, N) = (6.28, 0.628)$ in the region that satisfies (4.38). In fact, Fig. 4.2 indicates that if the positive T-periodic coefficient functions $a(t)$, $b(t)$ and $c(t)$ with $b(t) < c(t)$ are so chosen such that the corresponding M and N in (4.37) belong to the shaded region, this implies that the model (4.31) admits at least two positive T-periodic solutions.

In this section, we examined the existence of at least two positive T-periodic solutions for a scalar differential equation representing the dynamics of a renewable resource that is subjected to Allee effects. This study is physically relevant as it takes into account the seasonally dependent (cyclic) behavior in the intrinsic growth rate, Allee threshold, and carrying capacity for the renewable resource. While the

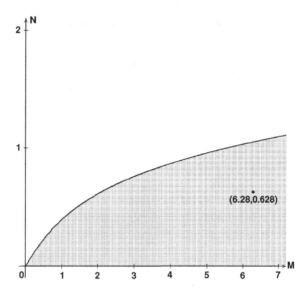

Fig. 4.2 The shaded region represents the portion in the (M, N) space that satisfies (4.38)

equation with constant coefficients (independent of strict periodicity) admits exactly two positive equilibrium solutions, the study undertaken in this section reveals that the equation with periodic coefficients admits at least two positive periodic solutions.

References

1. Bai, D., Xu, Y.: Periodic solutions of first order functional differential equations with periodic deviations. Comput. Math. Appl. **53**, 1361–1366 (2007)
2. Deimling, K.: Nonlinear Functional Analysis. Springer, Berlin (1985)
3. Freedman, H.I., Wu, J.: Periodic solutions of single species models with periodic delay. SIAM J. Math. Anal. **23**, 689–701 (1992)
4. Han, F., Wang, Q.: Existence of multiple positive periodic solutions for differential equation with state-dependent delays. J. Math. Anal. Appl. **324**, 908–920 (2006)
5. Jin, Z.L., Wang, H.: A note on positive periodic solutions of delayed differential equations. Appl. Math. Lett. **23**, 581–584 (2010)
6. Kuang, Y.: Delay Differential Equations with Applications in Population Dynamics. Academic Press, New York (1993)
7. Li, W.T., Fan, Y.H.: Existence and global attractivity of positive periodic solution for the impulsive delay Nicholson's blowflies model. J. Comput. Appl. Math. **201**, 55–68 (2007)
8. Liu, G., Yan, J., Zhang, F.: Existence of positive periodic solutions for neutral functional differential equations. Nonlinear Anal. **66**, 253–267 (2007)
9. Padhi, S., Srinivasu, P.D.N., Kumar, G.K.: Periodic solutions for an equation governing dynamics of a renewable resource subjected to allee effects. Nonlear. Anal.: Real World Appl. **11**, 2610–2618 (2010)
10. Padhi, S., Srivastava, S., Pati, S.: Three periodic solutions for a nonlinear first order functional differential equation. Appl. Math. Comput. **216**, 2450–2456 (2010)
11. Padhi, S., Srivsatava, S., Pati, S.: Positive periodic solutions for first order functional differential equations. Comm. Appl. Anal. **14**, 447–462 (2010)

12. Wang, H.: Positive periodic solutions of functional differential equations. J. Differ. Equ. **202**, 354–366 (2004)
13. Wang, Q., Dai, B.: Three periodic solutions of nonlinear neutral functional differential equations. Nonlinear Anal.: Real World Appl. **9**(3), 977–984 (2008)
14. Wu, Y.: Existence of positive periodic solutions for a functional differential equation with a parameter. Nonlinear Anal. **68**, 1954–1962 (2008)
15. Ye, D., Fan, M., Wang, H.: Periodic solutions for scalar functional differential equations. Nonlinear Anal. **62**, 1157–1181 (2005)
16. Zhang, W., Zhu, D., Bi, P.: Existence of periodic solutions of a scalar functional differential equation via a fixed point theorem. Math. Comput. Model. **46**, 718–729 (2007)

Chapter 5
Asymptotic Behavior of Periodic Solutions of Differential Equations of First Order

In this chapter[1], we establish results on the existence and global attractivity of the positive periodic solutions of mathematical models such as fishing models, blood cell production models, Nicholson's blowflies model, and the Lasota-Wazewska model.

Periodicity plays an important role in problems associated with real-world application such as those involving ecosystem dynamics and environmental variability. Periodic variation in the environment such as seasonal changes act upon the dynamic behavior of biological and ecological systems. Consequently, to analyze periodic variations in the environment, it is reasonable to consider the parameters involved to be periodic with same period.

Section 5.1 deals with sufficient conditions for the existence and global attractivity of a positive periodic solution of the fishing model

$$x'(t) = -a(t)x(t) + b(t)\frac{x(t)}{1+\left(\frac{x(t)}{p(t)}\right)^{\gamma}}, \tag{5.1}$$

where $a, b, p : [0, \infty) \to [0, \infty)$ are positive continuous periodic functions with period T, $x(t)$ is the population size at time t, $\frac{b(t)}{1+\left(\frac{x(t)}{p(t)}\right)^{\gamma}}$ is the birth rate, and $a(t)$ is the mortality rate. Here the parameter $\gamma > 0$ affects how density dependence sets in.

In Sect. 5.2, we provide some important results on the existence and global attractivity of positive periodic solutions of the Lasota-Wazewska model.

In Sect. 5.3, we investigate the global attractivity of a positive periodic solution of the red blood cell production model

$$x'(t) = -a(t)x(t) + b(t)\frac{x(t-\tau)}{1+x^n(t-\tau)}, \quad n > 0, \tag{5.2}$$

---

[1] Results in Sect. 5.1 are taken from [1–3, 5, 6, 9, 10, 14] and the results of Sect. 5.2 are taken from [7, 11, 12, 15, 16, 18, 25, 26]. The results given in Sects. 5.3 and 5.4 are new to the literature.

S. Padhi et al., *Periodic Solutions of First-Order Functional Differential Equations in Population Dynamics*, DOI: 10.1007/978-81-322-1895-1_5, © Springer India 2014

where $x(t)$ is the density of mature cells in blood circulation, $a(t)$ is the mortality rate, the function $b(t)\frac{x(t-\tau)}{1+x^n(t-\tau)}$ denotes the blood cell reproduction, and the time lag $x(t-\tau)$ describes the maturation phase before blood cells are released into circulation. Here, $a, b \in C(R_+, R_+)$ are periodic functions with period $T > 0$ and $\tau \in (0, \infty)$.

Section 5.4 is concerned with the global attractivity of a positive periodic solution of Nicholson's blowflies model

$$x'(t) = -a(t)x(t) + b(t)x(t-\tau)e^{-\gamma x(t-\tau)} \tag{5.3}$$

with constant delay. Here $\tau \in (0, \infty)$, $x(t)$ is the size of population at time t, $b(t)$ is the maximum per capita daily egg production, $\frac{1}{\gamma}$ is the size at which the population reproduces at its maximum rate, $a(t)$ is the per capita daily adult death rate, and τ is the generation time.

Examples and computer simulations using MATLAB are also given in each section to strengthen the given results.

For convenience, we introduce the notation that for any continuous T-periodic function $h : [0, \infty) \to R$, we set

$$h^* = \max_{0 \le t \le T} h(t) \quad \text{and} \quad h_* = \min_{0 \le t \le T} h(t).$$

5.1 Existence and Global Attractivity of Positive Periodic Solutions of Fishing Model

Differential equations of the form

$$x'(t) = [R(t, x) - M(t, x)]x - F(t)x, \tag{5.4}$$

where $R, M : [0, \infty) \times R \to R$ and $F : R \to R$ are continuous functions, are often used as population models, and in particular for models of fish populations (see, for example, [3, 9]). Here, $x(t)$ denotes the population size at time t, $R(t, x)$ is the birth rate, $M(t, x)$ is the mortality rate, and $F(t)$ denotes the harvesting rate. Seasonal effects including such things as weather, food supply, mating habits, and seasonal harvesting are often incorporated into these kind of models in the form of periodic coefficients. A common choice for the function R is often referred to as Hill's function [1–3, 9]

$$R(t, x) = \frac{b}{1 + \left(\frac{x}{p}\right)^\gamma},$$

where b and p are positive constants.

Berezansky and Idels [2] considered the delay differential equation model

$$x'(t) = \left[-a(t) + \frac{b(t)}{1 + \left(\frac{x(\theta(t))}{p(t)} \right)^\gamma} \right] x(t),$$ (5.5)

and they studied the existence of a periodic solution and its stability properties. In particular, they proved the existence of a positive periodic solution of (5.5) that is a global attractor for all other positive solutions of the equation. They also considered the case of Eq. (5.5) with proportional coefficients. Below, we summarize their results.

Suppose that

$$b(t) > a(t),$$ (5.6)

$$\sup_{t>0} \int_{\theta(t)}^{t} (b(s) - a(s)) \, ds < \infty, \quad \text{and} \quad \sup_{t>0} \int_{\theta(t)}^{t} a(s) \, ds < \infty.$$ (5.7)

Then there exists a global positive solution $x(t)$ to (5.5) together with an initial function

$$x(t) = \phi(t), \quad t < 0, \quad x(0) = x_0,$$ (5.8)

and this solution is persistent, that is, for the solution $x(t)$, there exist two constants α and β such that

$$0 < \alpha \le x(t) \le \beta < \infty,$$

where

$$\alpha = \min \left\{ x_0, \inf_{t \ge 0} p(t) \left(\frac{b(t)}{a(t)} - 1 \right)^{1/\gamma} \exp \left(-\sup_{t \ge 0} \int_{\theta(t)}^{t} a(s) \, ds \right) \right\}$$ (5.9)

and

$$\beta = \max \left\{ x_0, \sup_{t \ge 0} p(t) \left(\frac{b(t)}{a(t)} - 1 \right)^{1/\gamma} \exp \left(\sup_{t \ge 0} \int_{\theta(t)}^{t} (b(s) - a(s)) \, ds \right) \right\}.$$ (5.10)

Let $a(t), b(t), p(t)$ and $\theta(t)$ be T-periodic functions with $b(t) \ge a(t)$. If either

$$\inf_{t \ge 0} \left(\frac{b(t)}{a(t)} - 1 \right) p^\gamma(t) > 1$$ (5.11)

or

$$\sup_{t \geq 0} \left(\frac{b(t)}{a(t)} - 1 \right) p^{\gamma}(t) < 1, \tag{5.12}$$

then Eq. (5.5) has at least one positive solution $\overline{x}(t)$.
   If $0 \leq t - h(t) \leq \rho$, $q(t) \geq q_0 > 0$, and

$$\limsup_{t \to \infty} \int_{h(t)}^{t} q(s) \, ds < \frac{3}{2}, \tag{5.13}$$

then it is known (see [5, 10]) that every solution $x(t)$ of the first order delay differential equation

$$x'(t) + q(t)x(h(t)) = 0$$

satisfies the property

$$\lim_{t \to \infty} x(t) = 0.$$

This asymptotic property plays an important role in obtaining sufficient conditions for the global attractivity of solutions of many mathematical models in biology, as the following theorem shows.

Theorem 5.1.1 *Let $a(t)$, $b(t)$, $p(t)$ and $\theta(t)$ be periodic functions satisfying (5.6), (5.7), and either (5.11) or (5.12). In addition, assume that $p(t) \geq p > 0$,*

$$b(t) \geq b_0 > 0, \quad and \quad \gamma \int_{\theta(t)}^{t} b(s) \, ds < 6. \tag{5.14}$$

Then there exists an unique periodic solution $\overline{x}(t)$ of (5.5), and for every positive solution $x(t)$ of the system (5.5) and (5.8), we have

$$\lim_{t \to \infty} [x(t) - \overline{x}(t)] = 0,$$

that is, the positive periodic solution $\overline{x}(t)$ is a global attractor for all positive solutions to (5.5).

Proof The existence of a positive periodic solution $\overline{x}(t)$ of (5.5) follows from (5.6) and (5.11) or (5.12). If that solution is an attractor for all positive solutions of (5.5), then it is the unique positive periodic solution.

Set $x(t) = \exp(N(t))$ and rewrite (5.5) in the form

$$N'(t) = \frac{b(t)}{1 + \left(\frac{e^{N(\theta(t))}}{p(t)}\right)^{\gamma}} - a(t). \tag{5.15}$$

Suppose that $u(t)$ and $v(t)$ are two different solutions to (5.15). Let $w(t) = u(t) - v(t)$. To complete the proof of the theorem, it is sufficient to show that $\lim\limits_{t \to \infty} w(t) = 0$. Clearly, $w(t)$ is a solution of

$$w'(t) = b(t)\left[\frac{1}{1 + \left(\frac{e^{u(\theta(t))}}{p(t)}\right)^{\gamma}} - \frac{1}{1 + \left(\frac{e^{v(\theta(t))}}{p(t)}\right)^{\gamma}}\right]. \tag{5.16}$$

If we set

$$f(y, t) = \frac{1}{1 + \left(\frac{e^y}{p(t)}\right)^{\gamma}},$$

then by using the mean value theorem, we have for every t

$$f(y, t) - f(z, t) = f'(c)(y - z),$$

where

$$\min\{y, z\} \le c \le \max\{y, z\}.$$

Clearly,

$$f_y'(y, t) = -\frac{\gamma\left(\frac{e^y}{p(t)}\right)^{\gamma}}{\left[1 + \left(\frac{e^y}{p(t)}\right)^{\gamma}\right]^2}$$

and $|f_y'(y, t)| < \frac{1}{4}\gamma$. Then (5.16) takes the form

$$w'(t) = -M(t)w(\theta(t)), \tag{5.17}$$

where

$$M(t) = \frac{\gamma b(t)\left(\frac{e^{c(t)}}{p(t)}\right)^{\gamma}}{\left[1 + \left(\frac{e^{c(t)}}{p(t)}\right)^{\gamma}\right]^2},$$

and

$$\min\{u(\theta(t)), v(\theta(t))\} \le c(t) \le \max\{u(\theta(t)), v(\theta(t))\}.$$

Since $M(t) < \frac{1}{4}\gamma b(t)$, condition (5.13) is satisfied. Now, set $x_1(t) = e^{u(t)}$ and $x_2(t) = e^{v(t)}$, where $x_1(t)$ and $x_2(t)$ are two solutions to (5.5) corresponding to the solutions $u(t)$ and $v(t)$ of (5.15). Then,

$$M(t) \geq \frac{\gamma b_0 \left(\frac{\min\{\alpha,\beta\}}{p}\right)^{\gamma}}{\left(1 + \left(\frac{\min\{\alpha,\beta\}}{p}\right)^{\gamma}\right)^2} > 0,$$

where α and β are defined in (5.9) and (5.10). Hence, $M(t) \geq M_0 > 0$ for some constant $M_0 > 0$. Consequently, $\omega(t) \to 0$ as $t \to \infty$. This completes the proof of the theorem.

Consider Eq. (5.5) with proportional coefficients, i.e.,

$$x'(t) = \left[-a\,r(t) + \frac{b\,r(t)}{1 + \left(\frac{x(\theta(t))}{p}\right)^{\gamma}} \right] x(t), \tag{5.18}$$

where $r(t) \geq r_0 > 0$. Clearly, if $b > a$, then (5.18) has the unique positive equilibrium

$$x^* = \left(\frac{b}{a} - 1\right)^{1/\gamma} p. \tag{5.19}$$

Corollary 5.1.1 *If $b > a$, $\left(\frac{b}{a} - 1\right) p^{\gamma} \neq 1$, $r(t) \geq r_0 > 0$, and*

$$\gamma b \limsup_{t \to \infty} \int_{\theta(t)}^{t} r(s)\,ds < 6, \tag{5.20}$$

then the equilibrium x^* given in (5.19), is a global attractor for all positive solutions to (5.18).

Theorem 5.1.2 *Suppose $b > a$, $r(t) \geq r_0 > 0$, and*

$$\frac{\gamma(b - a)a}{b} \limsup_{t \to \infty} \int_{\theta(t)}^{t} r(s)\,ds < \frac{3}{2}. \tag{5.21}$$

Then the equilibrium $x^*(t)$ of (5.18), given in (5.19), is locally asymptotically stable.

Proof Setting $N(t) = x(t) - x^*(t)$, we obtain

$$N'(t) = \left[\frac{b\,r(t)}{1 + \left(\frac{N(\theta(t)) + x^*}{p}\right)^{\gamma}} - a\,r(t) \right] (N(t) + x^*(t)).$$

Letting

$$F(u(t), v(t)) = \left[\frac{b\,r(t)}{1 + \left(\frac{u + x^*}{p}\right)^{\gamma}} - a\,r(t) \right] (v(t) + x^*(t)),$$

we see that $F'_u(0,0) = -\frac{\gamma(b-a)a}{b}r(t)$ and $F'_v(0,0) = 0$. Hence, the linearized form of Eq. (5.18) is

$$N'(t) = -\frac{\gamma(b-a)a}{b}r(t)N(\theta(t)). \qquad (5.22)$$

Thus, in view of (5.21) and (5.13), it follows that (5.22) is asymptotically stable, and hence the positive equilibrium x^* of (5.18) is locally asymptotically stable. This completes the proof.   □

Now, we compare Theorems 5.1.1 and 5.1.2. We have $\max\{a(b-a)\} = b/4$. Therefore, if

$$b\gamma \limsup_{t\to\infty} \int_{\theta(t)}^{t} r(s)\,ds < 6, \qquad (5.23)$$

then (5.18) has a locally asymptotically stable equilibrium x^*.

The latter condition (5.23) does not depend on a, and is identical to condition (5.20) which guarantees the existence of a global attractor. Therefore, Theorem 5.1.2 provides the best possible conditions for the global attractivity for Eq. (5.5).

In another nice paper, Berezansky, Braverman and Idels [3] studied the population model

$$x'(t) = -a(t)x(t) + \frac{b(t)x(t)}{1+x^\gamma(t)} - r(t)x(\theta(t)), \qquad (5.24)$$

where $\frac{b(t)}{1+x^\gamma(t)}$ is a Hill's type function, $b(t)$ is the fecundity or birth rate, $a(t)$ is the mortality rate, $r(t)$ is the harvesting rate, γ is an abruptness rate, and all the coefficients are positive. In the following, we provide some results on the existence of global solutions for the initial value problem (5.24), (5.8), boundedness of these solutions, and extinction and persistence conditions. The following conditions are assumed while dealing with the initial value problem (5.24), (5.8):

(H_{43}) $\gamma > 0$;

(H_{44}) $b(t) \geq 0$, $a(t) \geq 0$, $r(t) \geq 0$ are Lebesgue measurable and essentially bounded functions on $[0, \infty)$, and $\lim_{t\to\infty} a(t) \geq a > 0$;

(H_{45}) $\theta(t)$ is a Lebesgue measurable function, $\theta(t) \leq t$, $\lim_{t\to\infty} \theta(t) = \infty$;

(H_{46}) $\phi : (-\infty, 0) \to R$ is a Borel measurable bounded function, $\phi(t) \geq 0$, and $x_0 > 0$.

We note that a locally absolutely continuous function $x : R \to R$ is called a solution of the initial value problem (5.24), (5.8), if it satisfies Eq. (5.24) for almost all $t \in [0, \infty)$ and (5.8) for $t \leq 0$.

Theorem 5.1.3 *Assume* (H_{43})–(H_{46}) *hold. Then the initial value problem* (5.24), (5.8) *has a unique local positive solution. This solution either becomes negative or is a global positive bounded solution.*

Theorem 5.1.3 justifies the well posedness of the initial value problem (5.24), (5.8). It has either a positive solution for all t which is bounded (the size of the population cannot grow infinitely due to the negative feedback, which is a typical situation in population dynamics) or it becomes negative in some finite time. If we apply Theorem 5.1.3 to the initial value problem

$$x'(t) = -A(t)x(t) + \frac{b(t)x(t)}{1 + x^{\gamma}(t)}, \quad x(0) = X_0 > 0, \tag{5.25}$$

we obtain the following corollary.

Corollary 5.1.2 *Assume* (H_{43})–(H_{46}) *hold with* $a(t)$ *replaced by* $A(t)$. *Then, there exists a unique positive bounded global solution of* (5.25).

Corollary 5.1.2 describes the source of possible extinction of a population. As long as there is no harvesting or there is no delay in the harvesting term, the solution is positive for all t.

Theorem 5.1.4 *Assume* (H_{43})–(H_{46}) *hold,*

$$0 \le \phi(t) \le x_0. \tag{5.26}$$

and, either

$$\sup_{t \ge 0} \int_{\theta(t)}^{t} r(s) \exp\left(-\int_{\theta(s)}^{s}\left[\frac{b(u)}{1 + A_0^{\gamma}} - a(u)\right] du\right) ds \le \frac{1}{e}, \tag{5.27}$$

where

$$A_0 = \sup_{t \ge 0} N(t) \tag{5.28}$$

with $N(t)$ *a solution of* (5.25), *or the inequality*

$$\sup_{t \ge 0} \int_{\theta(t)}^{t} [a(s) + r(s)] ds \le \frac{1}{e} \tag{5.29}$$

holds, with $X_0 = x_0$, *where either* $A(t) = a(t)$ *or* $A(t)$ *is denoted by*

$$A(t) = a(t) + r(t) \exp\left(-\int_{\theta(t)}^{t} [b(s) - a(s)] ds\right).$$

Then for the initial value problem (5.24), (5.8), *we have*

$$0 < x(t) \le A_0, \quad t \ge 0. \tag{5.30}$$

Theorem 5.1.4 gives sufficient conditions for the positiveness of solutions and provides an upper bound for solutions as well. Inequality (5.26) is vital for nonextinction in the following sense. If the harvesting rate is based on the size of the population at a previous time, then for the survival of the population, it is important that the field data on the population size is collected at the time when the population is not abundant. If the initial data is less than the initial function, then the harvesting based on the oversized estimation of the population can lead to early extinction (when the influence of the prehistory is still significant). In (5.27), sufficient condition on harvesting, mortality and growth rates, and the delay provide the solution is positive. The greater the mortality and harvesting rates are, the smaller should be delays providing that there is no extinction of the population in some finite time. Or else, for prescribed delays, a given natural growth rate $b(t)$, and the mortality rate $a(t)$, the harvesting rate should not exceed a certain number to avoid possible extinction. As the growth rate $b(t)$ becomes larger, the delay in the harvesting term may be larger.

Corollary 5.1.3 *Assume* (H_{43})–(H_{46}) *hold. Then,* (5.30) *holds for any positive solution of* (5.24).

Corollary 5.1.4 *Assume* (H_{43})–(H_{46}), (5.26) *and* (5.29) *hold. Then, the solution of* (5.24), (5.8) *is positive.*

Corollary 5.1.5 *Assume* (H_{43})–(H_{46}) *hold and* $b(t) = \alpha a(t)$, $\alpha > 0$. *Let* $x(t)$ *be a positive solution of* (5.24), (5.8).

(1) *If either* $\alpha \le 1$, *or* $\alpha > 1$ *and* $x_0 > x^* = (\alpha - 1)^{1/\gamma}$, *then* $x(t) \le x_0$.
(2) *If* $\alpha > 1$ *and* $x_0 \le x^*$, *then* $x(t) \le x^*$.

Now, consider the autonomous equation

$$x'(t) = \frac{bx(t)}{1 + x^\gamma(t)} - ax(t) - rx(t - \tau), \tag{5.31}$$

where $b > 0$, $a > 0$, $r > 0$, $\tau > 0$ and $\gamma > 0$.

Corollary 5.1.6 *Let* x^* *be denoted by*

$$x^* = \begin{cases} \left(\frac{b}{a} - 1\right)^{1/\gamma}, & a < b < a + re^{-\tau(b-a)}, \\ \left(\frac{b}{a+re^{-\tau(b-a)}} - 1\right)^{1/\gamma}, & b \ge a + re^{-\tau(b-a)}. \end{cases}$$

Then, for a positive solution of (5.31), (5.8), *we have*

$$x(t) \le A_0 = \max\{x_0, x^*\}. \tag{5.32}$$

If (5.26) *holds and either*

$$r\tau \exp\left\{-\left[\left(\frac{b}{1 + A_0^\gamma} - a\right)\tau - 1\right]\right\} \leq 1 \tag{5.33}$$

or

$$(a + r)\tau \leq 1/e,$$

where A_0 is denoted by (5.32), *then the solution of* (5.31), (5.8) *is positive.*

The estimate (5.30) illustrates an obvious fact that under harvesting, the solution cannot exceed what its value would be without harvesting. Corollary 5.1.5 deals with the situation when a nonconstant mortality rate is proportional to the birth rate. Corollary 5.1.6 states that at any point the solution does not exceed the maximum value of the equilibrium points without harvesting and the initial value.

We say that a solution $x(t)$ of (5.24), (5.8) is an *extinct solution*, if either

$$\lim_{t \to \infty} x(t) = 0,$$

or there exists $\bar{t} > 0$ such that $x(\bar{t}) = 0$. The bounded solution $x(t)$ is *persistent* if

$$\liminf_{t \to \infty} x(t) > 0.$$

Theorem 5.1.5 *Assume that* (H_{43})–(H_{46}) *hold,*

$$\int_0^\infty [a(t) + r(t) - b(t)]\,dt = \infty$$

and either

(i) $a(t) \geq b(t)$, *or*
(ii) $a(t) + r(t) \geq b(t) \geq a(t)$ *and* $\limsup_{t \to \infty}[b(t) - a(t)](t - \theta(t)) < 1$.

Then, every solution of (5.24), (5.8) *is an extinct one.*

Applying Theorem 5.1.5 to the autonomous Eq. (5.31), we obtain the following corollary.

Corollary 5.1.7 *Assume* (i) $a \geq b$ *or* (ii) $a + r \geq b \geq a$ *and* $(b - a)\tau < 1$. *Then, every solution of Eq.* (5.31) *is an extinct one.*

Theorem 5.1.5 shows that if the total of the mortality and harvesting rates exceeds the birth rate, then the population is destined for extinction. That is, the population either equals zero at some finite time or tends to zero as $t \to \infty$.

Theorem 5.1.6 *Assume the conditions of Theorem 5.1.4 hold and the solution of the problem*

$$x'(t) = \frac{b(t)x(t)}{1 + x^\gamma(t)} - a(t)x(t) - r(t)A_0, \quad x(0) = x_0$$

is positive and persistent, where A_0 is denoted by (5.28). Then the solution of (5.24), (5.8) is persistent.

Theorem 5.1.6 provides sufficient conditions for the persistence of solutions. Sometimes the existence of the lower bound for solutions is not less important than positiveness of solutions. For instance, for blood diseases, the mortality rate of white or red blood cells is greater than zero. This means that for smaller values of x, the model can become irrelevant.

Now, we consider Eq. (5.1) under conditions different from those in [2, 3]. In the remainder of this section, we assume that $\gamma > 1$.

The following Lemma shows that given a positive initial condition, solutions of Eq. (5.1) are positive and bounded.

Lemma 5.1.1 *Every solution of (5.1) having positive initial condition is positive and satisfies the property*

$$\limsup_{t \to \infty} x(t) < K, \tag{5.34}$$

where

$$K = \frac{b^*}{a_*} p_* \left(\frac{p^*}{p_*}\right)^\gamma \left(\frac{1}{\gamma}\right) (\gamma - 1)^{\frac{\gamma-1}{\gamma}}. \tag{5.35}$$

Proof Let $x(t)$ be a solution of (5.1) with $x(0) = x_0 > 0$; then

$$x(t) = x_0 \exp \int_0^t \left[-a(s) + \frac{b(s)}{1 + (\frac{x(s)}{p(s)})^\gamma} \right] ds.$$

Hence, $x(t)$ is defined on $[0, \infty)$ and $x(t) > 0$ for $t \geq 0$.

Next, we claim that $x(t)$ is bounded. Setting

$$\tilde{g}(x) = \frac{p^{*\gamma} x(t)}{p_*^\gamma + x^\gamma(t)},$$

we see that

$$\tilde{g}(x) \leq \bar{g} = \frac{p^{*\gamma} \mu}{p_*^\gamma + \mu^\gamma} \quad \text{where } \mu = p_* \left(\frac{1}{\gamma - 1}\right)^{\frac{1}{\gamma}}. \tag{5.36}$$

From (5.1), we obtain

$$x'(t) \leq -a(t)x(t) + b^*\bar{g}. \tag{5.37}$$

To prove (5.34), note that from (5.36) and (5.37), we have

$$
\begin{aligned}
x(t) &\leq x_0 e^{-\int\limits_0^t a(s)\,ds} + \int\limits_0^t b^*\bar{g}\, e^{-\int\limits_s^t a(u)\,du}\,ds \\
&\leq x_0 e^{-a_* t} + b^*\bar{g}\int\limits_0^t e^{-a_*(t-s)}\,ds \\
&\leq x_0 e^{-a_* t} + \frac{b^*\bar{g}}{a_*}(1 - e^{-a_* t}).
\end{aligned}
$$

This in turn implies that

$$
\limsup_{t\to\infty} x(t) \leq \frac{b^*\bar{g}}{a_*} = K,
$$

where K is as given in (5.35). This completes the proof of the lemma. $\square$

Remark 5.1.1 Note that from the proof of Lemma 5.1.1, we have

$$
x(t) < x_0 + \frac{b^*\bar{g}}{a_*} = x_0 + K
$$

for all $t \geq 0$.

The following theorem gives a sufficient condition for the existence of a positive periodic solution of (5.1).

Theorem 5.1.7 *If $b_* > a^*$, then (5.1) has at least one positive periodic solution.*

Proof From Lemma 5.1.1, it follows that every solution of (5.1) with a positive initial condition is positive and bounded. Furthermore, the function $\tilde{g}(x) = \frac{p^{*\gamma} x(t)}{p^{*\gamma} + x^{\gamma}(t)}$ is decreasing in $(p^*\left(\frac{1}{\gamma-1}\right)^{\frac{1}{\gamma}}, \infty)$. Consider the function

$$
f(x) = -ax + b\frac{x}{1 + \left(\frac{x}{p^*}\right)^{\gamma}},
$$

where a and b are constants. Clearly, the function attains its maximum

$$
\begin{aligned}
f_{\max} = {}&\frac{p^*[(4ab\gamma + (\gamma-1)^2 b^2)^{\frac{1}{2}} - (2a + (\gamma-1)b)]^{\frac{1}{\gamma}}}{(2a)^{\frac{1}{\gamma}}} \\
&\times\left[-a + \frac{(4ab\gamma + (\gamma-1)^2 b^2)^{\frac{1}{2}} + (\gamma-1)b}{2\gamma}\right]
\end{aligned}
$$

at

$$x = \hat{x} = \frac{p^*}{(2a)^{\frac{1}{\gamma}}}[(4ab\gamma + (\gamma - 1)^2 b^2)^{\frac{1}{2}} - (2a + (\gamma - 1)b)]^{\frac{1}{\gamma}}.$$

Now $\hat{x} > 0$ and $f_{\max} > 0$ for $b > a$, and since $f(x) \to -\infty$ as $x \to \infty$, it follows that there exists an $\alpha \in (\hat{x}, \infty)$ such that $f(\alpha) = 0$. Thus, $f_1(x) = 0$ and $f_2(x) = 0$ have roots x_1 and x_2 respectively, where

$$f_1(x) = \frac{b_* x}{1 + \left(\frac{x}{p^*}\right)^{\gamma}} - a^* x$$

and

$$f_2(x) = \frac{b^* x}{1 + \left(\frac{x}{p^*}\right)^{\gamma}} - a_* x.$$

A simple calculation also shows that

$$\max\left\{\hat{x}, p^*\left(\frac{1}{\gamma - 1}\right)^{\frac{1}{\gamma}}\right\} < x_1 < x_2.$$

Let $x(t) = x(t, 0, \alpha)$, $\alpha \geq \max\left\{\hat{x}, p^*\left(\frac{1}{\gamma-1}\right)^{\frac{1}{\gamma}}\right\}$, be the unique solution of (5.1) through $(0, \alpha)$. We claim that $x(t) \in [x_1, x_2]$ for $t \geq 0$. Suppose this is not the case; say, let

$$t_1 = \inf\{t > 0 : x(t) > x_2\}.$$

Then there exists a $t_2 \geq t_1$ such that $x(t_2) > x_2$ and $x'(t_2) > 0$. Hence, from (5.1) we obtain

$$0 < x'(t_2) = -a(t_2)x(t_2) + \frac{b(t_2)x(t_2)}{1 + \left(\frac{x(t_2)}{p(t_2)}\right)^{\gamma}}$$

$$\leq -a_* x_2 + \frac{b^* x_2}{1 + \left(\frac{x_2}{p^*}\right)^{\gamma}} = f_2(x_2) = 0,$$

which is a contradiction. Consequently, $x(t) \leq x_2$. By a similar argument, we can show that $x(t) \geq x_1$. Thus, in particular,

$$x_T = x(T, 0, \alpha) \in [x_1, x_2].$$

Next, we define a mapping $F : [x_1, x_2] \to [x_1, x_2]$ as follows: for each $\alpha \in [x_1, x_2]$, let $F(\alpha) = x_T$. Since the solution $x(t, 0, \alpha)$ depends continuously on the initial value α, the mapping F is continuous and maps the interval $[x_1, x_2]$ into

$[x_1, x_2]$. By Brouwer's fixed point theorem, F has a fixed point $\bar{x}$. Thus, the unique solution $\bar{x} = x(t, 0, \alpha)$ is periodic with period T. This completes the proof of the theorem.   □

Remark 5.1.2 We want to show that $\bar{x}(t)$ is an attractor to all other positive solutions of (5.1). To prove the attractivity theorems, we need the following lemma. An indirect proof of this Lemma can be found in [14]. We present a proof here for the sake of completeness.

Definition 5.1.1 A function $x \in C([0, \infty), R)$ is said to be nonoscillatory, if there exists a $t_1 \geq 0$ such that $x(t) > 0$ or < 0 for $t \geq t_1$; otherwise, $x(t)$ is called oscillatory. By a solution of any of the above-mentioned differential equations, we mean a real-valued solution of that equation which exists on $[t_0, \infty)$ and is nontrivial in any neighborhood of infinity, where $t_0 \geq 0$ depends on the solution.

Lemma 5.1.2 *Let $z \in C^1([0, \infty), R)$ and $\sigma \in R$. If $z \in L^2[\sigma, \infty)$ and $z'(t)$ is bounded, then $z(t) \to 0$ as $t \to \infty$.*

Proof Clearly, the statement holds for any nonoscillatory function on $[0, \infty)$. So let $z(t)$ be oscillatory on $[0, \infty)$. Since $z \in L^2[0, \infty)$, $\int_0^\infty z^2(t)dt < \infty$. This in turn implies that $\liminf_{t\to\infty} z(t) = 0$. To complete the proof of the theorem, it remains to show that $\limsup_{t\to\infty} z(t) = 0$. Suppose that $\limsup_{t\to\infty} z(t) \neq 0$. Then there exist an $\epsilon > 0$ and a sequence $\{t_n\}_{n=1}^\infty$ such that $t_n \to \infty$ as $n \to \infty$ and $z(t_n) > 2\epsilon$, for large n. Since $\liminf_{t\to\infty} z(t) = 0$, there exists a sequence $\{t_n^*\}_{n=1}^\infty$ such that $t_n^* \to \infty$ as $n \to \infty$ and $z(t_n^*) \to 0$ as $n \to \infty$. Thus $z(t_n^*) < \epsilon$ for $n \geq N_\epsilon$ for some positive integer N_ϵ. It is possible to extract sequences $\{s_n\}_{n=1}^\infty$ and $\{\sigma_n\}_{n=1}^\infty$ such that $\sigma_n < s_n < \sigma_{n+1}$, $\sigma_n \to \infty$ as $n \to \infty$, $z(s_n) > 2\epsilon$ and $z(\sigma_n) < \epsilon$ for $n \geq N_\epsilon$. Since $z(t)$ is continuous, there exist sequences $\{\tau_n\}_{n=1}^\infty$ and $\{\tau_n^*\}_{n=1}^\infty$ such that $\sigma_n < \tau_n^* < \tau_n < s_n$ with $z(\tau_n^*) = \epsilon$ and $z(\tau_n) = 2\epsilon$. It is clear that the intervals (τ_n^*, τ_n) are disjoint. Then

$$\sum_{n=1}^\infty (\tau_n - \tau_n^*)\epsilon^2 \leq \sum_{n=1}^\infty \int_{\tau_n^*}^{\tau_n} z^2(t)\, dt \leq \int_0^\infty z^2(t)\, dt < \infty,$$

which implies that $\lim_{n\to\infty}(\tau_n - \tau_n^*) = 0$. By the mean value theorem, the calculation

$$z'(\xi_n) = \frac{z(\tau_n) - z(\tau_n^*)}{\tau_n - \tau_n^*}, \quad \tau_n^* < \xi_n < \tau_n,$$

implies that $z'(\xi_n) \to \infty$ as $n \to \infty$, which contradicts the hypothesis of the lemma. Hence our claim holds, that is, $\limsup_{t\to\infty} z(t) = 0$. Consequently, $z(t) \to 0$ as $t \to \infty$, and the lemma is proved.   □

Theorem 5.1.8 *Assume that*

(H_{47}) $b^* p^{*\gamma}(\gamma - 1)^2 < 4a_*(p^{*\gamma} + p_*^\gamma(\gamma - 1))$.

Then $\bar{x}(t)$ is a global attractor to all other positive solutions of (5.1), that is, every positive solution $x(t)$ of (5.1) satisfies

$$\lim_{t\to\infty} [x(t) - \bar{x}(t)] = 0.$$

Proof Setting $Z(t) = x(t) - \bar{x}(t)$, we obtain

$$Z'(t) = -a(t)Z(t) + b(t)\left[\frac{x(t)}{1 + (\frac{x(t)}{p(t)})^\gamma} - \frac{\bar{x}(t)}{1 + (\frac{\bar{x}(t)}{p(t)})^\gamma}\right]. \qquad (5.38)$$

Equation (5.38) is equivalent to

$$\left(\frac{1}{2}Z^2(t)\right)' = -a(t)Z^2(t) + Z(t)b(t)p^\gamma(t)\left[\frac{x(t)}{p^\gamma(t) + x^\gamma(t)} - \frac{\bar{x}(t)}{p^\gamma(t) + \bar{x}^\gamma(t)}\right]. \qquad (5.39)$$

Let $F(t, \theta) = \frac{\theta}{p^\gamma(t) + \theta^\gamma}$; then

$$\frac{\partial}{\partial\theta}F(t, \theta) = \frac{p^\gamma(t) + (1 - \gamma)\theta^\gamma}{(p^\gamma(t) + \theta^\gamma)^2} < F_1(t, \theta) = \frac{p^{*\gamma} + (1 - \gamma)\theta^\gamma}{(p^\gamma(t) + \theta^\gamma)^2},$$

where θ lies between $x(t)$ and $\bar{x}(t)$. Since $F_1(t, \theta) < 0$ for $\theta > \frac{p^*}{(\gamma-1)^{\frac{1}{\gamma}}}$, set

$$G_1(t, \theta) = -F_1(t, \theta) = \frac{(\gamma - 1)\theta^\gamma - p^{*\gamma}}{(p^\gamma(t) + \theta^\gamma)^2}.$$

Note that $G_1(t, \theta) > 0$ for $\theta > \frac{p^*}{(\gamma-1)^{\frac{1}{\gamma}}}$. Furthermore, $G_1(t, \theta) < G(\theta) = \frac{(\gamma-1)\theta^\gamma - p^{*\gamma}}{(p_*^\gamma + \theta^\gamma)^2}$. A simple calculation shows that $G(\theta)$ attains its maximum value

$$\frac{(\gamma - 1)^2}{4(p_*^\gamma(\gamma - 1) + p^{*\gamma})}$$

at

$$\theta = \left[\frac{2p^{*\gamma} + (\gamma - 1)p_*^\gamma}{\gamma - 1}\right]^{\frac{1}{\gamma}}.$$

Applying the mean value theorem, (5.39) yields

$$\left(\frac{1}{2}Z^2(t)\right)' \leq -a(t)Z^2(t) + Z(t)b(t)p^\gamma(t)|x(t) - \overline{x}(t)|G(\theta)$$

$$\leq -a(t)Z^2(t) + b(t)p^\gamma(t)G(\theta)Z^2(t)$$

$$\leq -\left[a_* - b^*p^* \frac{(\gamma - 1)^2}{4(p_*^\gamma(\gamma - 1) + p^{*\gamma})}\right]Z^2(t),$$

that is,

$$\left(\frac{1}{2}Z^2(t)\right)' \leq -\mu Z^2(t), \tag{5.40}$$

where

$$\mu = a_* - b^*p^{*\gamma} \frac{(\gamma - 1)^2}{4(p_*^\gamma(\gamma - 1) + p^{*\gamma})} > 0$$

by (H_{47}). Integrating inequality (5.40) from t_1 to t with $t_1 \geq \frac{p^*}{(\gamma - 1)^{\frac{1}{\gamma}}}$, we obtain

$$\mu \int_{t_1}^{t} Z^2(s)\,ds \leq \frac{1}{2}Z^2(t_1) - \frac{1}{2}Z^2(t) \leq \frac{1}{2}Z^2(t_1).$$

This shows $\int_{t_1}^\infty Z^2(s)\,ds < \infty$, i.e., $Z \in L^2[t_1, \infty)$. Furthermore, the boundedness of $x(t)$ and $\overline{x}(t)$ imply that $Z'(t)$ is bounded. Hence, by Lemma 5.1.2, $Z(t) \to 0$ as $t \to \infty$, that is, $x(t) \to \overline{x}(t)$ as $t \to \infty$. This completes the proof of the theorem. $\square$

The next theorem gives another sufficient condition for $\overline{x}(t)$ to be a global attractor to all other positive solutions of (5.1). As we shall see in the examples given later, this condition is independent of the one given in Theorem 5.1.8.

Theorem 5.1.9 *Suppose that*

$$(H_{48}) \quad a_* > b^*p^{*\gamma}\left[\frac{(\gamma - 1)K^\gamma}{p_*^{2\gamma}} - \frac{p^{*\gamma}}{(p^{*\gamma} + K^\gamma)^2}\right].$$

Then $\overline{x}(t)$ is global attractor to all other positive solutions of (5.1).

Proof Letting $Z(t) = x(t) - \overline{x}(t)$ and proceeding as in the proof of the Theorem 5.1.8, we again obtain (5.39). Applying the mean value theorem we have

$$\left(\frac{1}{2}Z^2(t)\right)' \leq -a(t)Z^2(t) + Z(t)b(t)p^\gamma(t)|x(t) - \overline{x}(t)|G_1(t, \theta). \tag{5.41}$$

Then,

$$G_1(t, \theta) = \frac{(\gamma - 1)\theta^\gamma - p^{*\gamma}}{(p^\gamma(t) + \theta^\gamma)^2} = \frac{(\gamma - 1)\theta^\gamma}{(p^\gamma(t) + \theta^\gamma)^2} - \frac{p^{*\gamma}}{(p^\gamma(t) + \theta^\gamma)^2}$$

$$< \frac{(\gamma - 1)\theta^\gamma}{p^{2\gamma}(t)} - \frac{p^{*\gamma}}{(p^\gamma(t) + \theta^\gamma)^2}$$

$$\leq \frac{(\gamma - 1)K^\gamma}{p_*^{2\gamma}} - \frac{p^{*\gamma}}{(p^{*\gamma} + K^\gamma)^2}.$$

Since $G_1(t, \theta) > 0$ for $\theta > \frac{p^*}{(\gamma - 1)^{\frac{1}{\gamma}}}$, we have $\frac{(\gamma - 1)K^\gamma}{p_*^{2\gamma}} - \frac{p^{*\gamma}}{(p^{*\gamma} + K^\gamma)^2} > 0$ for $\theta > \frac{p^*}{(\gamma - 1)^{\frac{1}{\gamma}}}$.

Hence, from (5.41), we obtain

$$\left(\frac{1}{2}Z^2(t)\right)' \leq \left[-a(t) + b(t)p^\gamma(t)\left(\frac{(\gamma - 1)K^\gamma}{p_*^{2\gamma}} - \frac{p^{*\gamma}}{(p^{*\gamma} + K^\gamma)^2}\right)\right]Z^2(t)$$

$$\leq -\left[a_* - b^* p^{*\gamma}\left(\frac{(\gamma - 1)K^\gamma}{p_*^{2\gamma}} - \frac{p^{*\gamma}}{(p^{*\gamma} + K^\gamma)^2}\right)\right]Z^2(t),$$

or

$$\left(\frac{1}{2}Z^2(t)\right)' \leq -\mu Z^2(t),$$

where

$$\mu = a_* - b^* p^{*\gamma}\left(\frac{(\gamma - 1)K^\gamma}{p_*^{2\gamma}} - \frac{p^{*\gamma}}{(p^{*\gamma} + K^\gamma)^2}\right) > 0$$

by (H_{48}). The remainder of the proof of the theorem is similar to the proof of the Theorem 5.1.8.   □

Remark 5.1.3 A somewhat easier condition to verify than (H_{48}) is

(H_{49}) $\left(\frac{b^*}{a_*}\right)^{\gamma+1} \left(\frac{p^*}{p_*}\right)^{\gamma(\gamma+1)} \left(\frac{\gamma-1}{\gamma}\right)^\gamma < 1.$

It can be shown that this condition implies (H_{48}).

Example 5.1.1 Consider the equation

$$x'(t) = -\left(1 + \frac{\sin^2 t}{10}\right)x(t) + \frac{\left(1.5 + \frac{\cos^2 t}{20}\right)x(t)}{1 + \left(\frac{x(t)}{2+\cos^2 t}\right)^3}, \quad t \geq 0.$$

Here $a(t) = 1 + \frac{\sin^2 t}{10}$, $b(t) = 1.5 + \frac{\cos^2 t}{20}$, $p(t) = 2 + \cos^2 t$ and $\gamma = 3$. Clearly $b_* = 1.5 > 1.1 = a^*$. It is easy to see that the condition (H_{47}) of Theorem 5.1.8 is satisfied. Therefore, this equation has a positive periodic solution that is global attractor to all other positive solutions. On the other hand

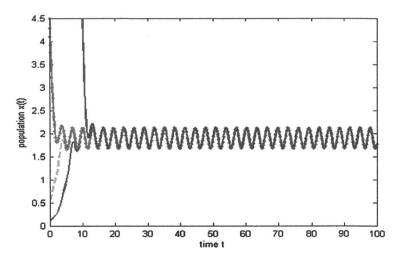

Fig. 5.1 Computer simulation of Example 5.1.1

$$b^* p^{*\gamma} \left[\frac{(\gamma - 1)K^\gamma}{p_*^{2\gamma}} - \frac{p^{*\gamma}}{(p^{*\gamma} + K^\gamma)^2} \right] = 221.87 > 1 = a_*$$

implies that (H_{48}) fails to hold. Consequently, Theorem 5.1.9 cannot be applied to this example (Fig. 5.1).

Example 5.1.2 Consider the equation

$$x'(t) = - \left(1.2 + \frac{\sin^2 t}{50} \right) x(t) + \frac{\left(1.3 + \frac{\cos^2 t}{100} \right) x(t)}{1 + \left(\frac{x(t)}{0.5 + \frac{\sin^2 t}{1000}} \right)^6}, \quad t \geq 0.$$

Here $a(t) = 1.2 + \frac{\sin^2 t}{50}$, $b(t) = 1.3 + \frac{\cos^2 t}{100}$, $p(t) = 0.5 + \frac{\sin^2 t}{1000}$, and $\gamma = 6$.
Now $b_* = 1.3 > 1.22 = a^*$, and

$$\left(\frac{b^*}{a_*} \right)^{\gamma+1} \left(\frac{p^*}{p_*} \right)^{\gamma(\gamma+1)} \left(\frac{\gamma - 1}{\gamma} \right)^\gamma = 0.672956783 < 1,$$

that is, (H_{49}) and hence (H_{48}) is satisfied. Thus, Theorem 5.1.9 can be applied and so the equation has a positive periodic solution that is a global attractor to all other positive solutions. On the other hand, the condition (H_{47}) fails to hold, which means that Theorem 5.1.8 cannot be applied to this example (Fig. 5.2).

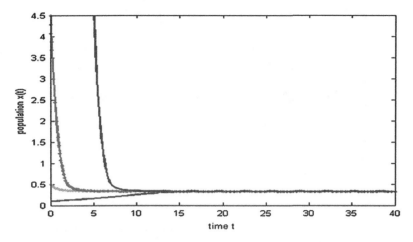

Fig. 5.2 Computer simulation of Example 5.1.2

Example 5.1.3 Consider the equation

$$x'(t) = -\left(1.5 + \frac{\sin^2 t}{10}\right)x(t) + \frac{\left(1.7 + \frac{\cos^2 t}{20}\right)x(t)}{1 + \left(\frac{x(t)}{4 + \frac{\sin^2 t}{100}}\right)^6}, \quad t \geq 0.$$

Here $a(t) = 1.5 + \frac{\sin^2 t}{10}$, $b(t) = 1.7 + \frac{\cos^2 t}{20}$, $p(t) = 4 + \frac{\sin^2 t}{100}$, $\gamma = 6$. Since $b_* = 1.7 > 1.6 = a^*$, $K = 3.0189$, and

$$b^* p^{*\gamma}\left[\frac{(\gamma - 1)K^\gamma}{p_*^{2\gamma}} - \frac{p^{*\gamma}}{(p^{*\gamma} + K^\gamma)^2}\right]$$

$$= (1.75)(4.01)^6\left[\frac{5(3.0189)^6}{4^{12}} - \frac{4.01^6}{(4.01^6 + 3.0189^6)^2}\right]$$

$$= 0.38908 < 1.5 = a_*.$$

Condition (H_{48}) of Theorem 5.1.9 is satisfied. This shows that again we have a positive periodic solution that is a global attractor to all other positive solutions. However, a simple calculation shows that (H_{47}) fails to hold, so Theorem 5.1.8 cannot be applied to this example (Fig. 5.3).

The final example in this section is a simple one, yet it shows that Theorem 5.1.8 may hold when those in [2] do not.

Example 5.1.4 Consider the equation

$$x'(t) = -x(t) + \frac{2x(t)}{1 + x^2(t)}, \quad t \geq 0.$$

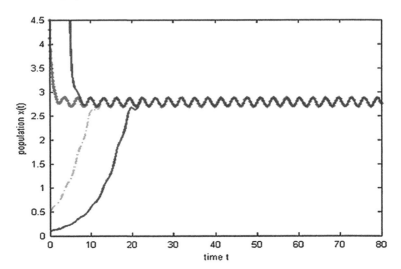

Fig. 5.3 Computer simulation of Example 5.1.3

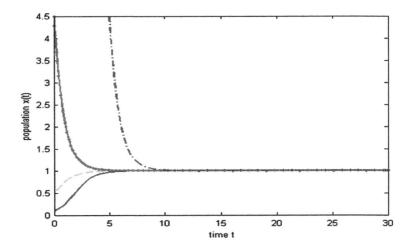

Fig. 5.4 Computer simulation of Example 5.1.4

Here, $a(t) = 1 < b(t) = 2$, $p(t) = 1$ and $\gamma = 2$. It is easy to see that condition (H_{47}) of Theorem 5.1.8 is satisfied, so this equation has a positive periodic solution globally attracting all other positive solutions. In fact, this solution is $x(t) \equiv 1$. However, Theorem 2.1 in [2] does not apply to this example (Fig. 5.4).

5.2 Existence and Global Attractivity of Positive Periodic Solutions of Lasota-Wazewska Model

The nonlinear delay differential equation

$$y'(t) = -ay(t) + be^{-\gamma y(t-\tau)}, \quad t \geq 0, \tag{5.42}$$

where a, b, γ, and $\tau \in (0, \infty)$, has been proposed by Wazewska-Czyzewska and Lasota [26] as a model for the survival of red blood cells in an animal. Here $y(t)$ denotes the number of blood cells at time t, a is the probability of death of red blood cells, b and γ are positive constants related to the productions of blood cells per unit time, and τ is the time required to produce a red blood cell.

Equation (5.42) is a particular case of the nonlinear delay differential equation

$$x'(t) = -a(t)x(t) + b(t)e^{-\gamma(t)x(t-\tau(t))}, \quad t \geq 0, \tag{5.43}$$

where $a(t)$, $b(t)$, $\gamma(t)$, and $\tau(t)$ are positive and continuous functions. Taking $a(t) \equiv a$, $b(t) \equiv b$, $\gamma(t) \equiv \gamma$, and $\tau(t) \equiv \tau$ reduces Eq. (5.43) to (5.42).

Since we are concerned with positive periodic solutions, we assume that the functions $a(t)$, $b(t)$, $\gamma(t)$, and $\tau(t)$ are positive T-periodic functions, and $T > 0$ is a real number. Together with (5.43), we consider the initial condition

$$x(t) = \phi(t), \quad -\tau \leq t \leq 0, \ \phi(0) > 0, \ \phi \in C([-\tau, 0], R^+), \tag{5.44}$$

where $\tau = \max_{0 \leq t \leq T} \tau(t)$. Then by the method of steps (see Driver [4]), it follows that the initial value problem (5.43)–(5.44) has a unique positive solution $x(t)$ for all $t \geq 0$.

In the following, we first prove some results on the existence of a positive periodic solution $x^*(t)$ of (5.43) using a classical fixed point theorem. Then we show that all solutions of (5.43) tend to $x^*(t)$ eventually.

First suppose that $\gamma(t) \equiv 1$ and $\tau(t) = mT$, where m is a nonnegative integer. Then Eq. (5.43) can be expressed as

$$x'(t) = -a(t)x(t) + b(t)e^{-\gamma(t)x(t-mT)}, \quad t \geq 0. \tag{5.45}$$

Equation (5.42) is a particular case of (5.45). In fact, the transformation $a(t) \equiv a$, $b(t) \equiv b\gamma$ with a, b, and γ positive constants, transforms Eq. (5.45) into (5.42) with $y(t) = x(t)/\gamma$.

Theorem 5.2.1 *Equation (5.45) has a positive periodic solution $x^*(t)$ with period T.*

Proof First, consider Eq. (5.45) without a delay, that is, we consider the equation

$$x'(t) = -a(t)x(t) + b(t)e^{-x(t)}, \quad t \geq 0. \tag{5.46}$$

Observe that there is a unique $r(t) > 0$ such that

$$-a(t)r(t) + b(t)e^{-r(t)} = 0 \text{ for } t \geq 0.$$

Set

$$A = \min_{0 \leq t \leq T} r(t) \quad \text{and} \quad B = \max_{0 \leq t \leq T} r(t),$$

and let $x(t) = x(t, 0, x_0)$ denote the unique solution of (5.46) through $(0, x_0)$.

We claim that $x_0 \in [A, B]$ implies that $x(t) = x(t, 0, x_0) \in [A, B]$. First, we show that $x(t) \leq B$. If this is not the case, there exists

$$t^* = \inf\{t \geq 0 : x(t) > B\} < \infty.$$

Then, it is easy to see that there exists a $t_1 > t^*$ such that

$$x(t_1) > B \text{ and } x'(t_1) > 0.$$

Hence, it follows from (5.46), that

$$\begin{aligned}
0 < x'(t_1) &= -a(t_1)x(t_1) + b(t_1)e^{-x(t_1)} \\
&\leq -a(t_1)B + b(t_1)e^{-B} \\
&\leq -a(t_1)r(t_1) + b(t_1)e^{-r(t_1)} = 0,
\end{aligned}$$

which is a contradiction. Hence, $x(t) \leq B$. By a similar argument, we can show that $x(t) \geq A$. Thus,

$$x_T = x(T, 0, x_0) \in [A, B].$$

Now, define a mapping $F : [A, B] \to [A, B]$ as follows: For each $x_0 \in [A, B]$, let

$$F(x_0) = x_T.$$

Since the function $x(t, 0, x_0)$ depends continuously on the initial value x_0, the mapping F is continuous and maps the interval $[A, B]$ into itself. Therefore, F has a fixed point x_0^* by Brouwer's fixed point theorem. Thus, the unique positive solution $x^*(t) = x(t, 0, x^*)$ is periodic with period T.

Finally, by noting that $x^*(t) = x^*(t - mT)$, we see that $x^*(t)$ is also a positive T-periodic solution of (5.45). This completes the proof.   □

Remark 5.2.1 Let $\gamma(t) \equiv \gamma > 0$ be a constant. Let $x(t)$ be any other positive solution of (5.46). Assume that $x(t) < x^*(t)$ for t sufficiently large (the case when $x(t) > x^*(t)$ can be treated similarly). Set

$$x(t) = x^*(t) - \frac{1}{\gamma}y(t);$$

then $y(t) > 0$ for t sufficiently large, and $y(t)$ satisfies the equation

$$y'(t) + a(t)y(t) + \gamma b(t)e^{-\gamma x^*(t)}\left[1 - e^{-y(t)}\right] = 0.$$

However, since $y(t) > 0$ for t sufficiently large, it follows that

$$y'(t) \leq -\gamma b(t)e^{-\gamma x^*(t)}\left[1 - e^{-y(t)}\right] < 0.$$

Thus, $y(t)$ is decreasing, and therefore $\lim_{t\to\infty} y(t) = \mu \in [0, \infty)$. We now show that $\mu = 0$.

If $\mu > 0$, then there exists an $\epsilon > 0$ and $t_\epsilon > 0$ such that for $t \geq t_\epsilon, 0 < \mu - \epsilon < y(t) < \mu + \epsilon$. On the other hand, the above inequality gives

$$y'(t) \leq -\gamma b(t)e^{-\gamma x^*(t)}\left[1 - e^{-\gamma(\mu-\epsilon)}\right], \quad t \geq t_\epsilon.$$

An integration for t_ϵ yields a contradiction to the fact that $y(t) > 0$ for t sufficiently large. Thus, we have

$$\lim_{t\to\infty}\left[x(t) - x^*(t)\right] = \lim_{t\to\infty}\frac{1}{\gamma}y(t) = 0.$$

We then have the following corollary.

Corollary 5.2.1 *Let $\gamma(t) \equiv \gamma$ be a positive constant. Then:*

(i) *There is a unique T-periodic solution $x^*(t)$ of (5.46).*
(ii) *For every positive solution $x(t)$ of (5.46), we have*

$$\lim_{t\to\infty}\left[x(t) - x^*(t)\right] = 0.$$

From the proof of Theorem 5.2.1, we see that $x^*(t)$ satisfies $A \leq x^*(t) \leq B$, which gives an estimate for the location of the periodic solution. Clearly, when $a(t)$ and $b(t)$ are constants, then $A = B$ and so $x^*(t) \equiv A$.

Lemma 5.2.1 *Consider the difference equation*

$$A_{n+1} = h(A_n), \tag{5.47}$$

where $h \in C^1[R, R]$. Assume that h is a nonincreasing function and has a unique fixed point A^*. Suppose also that

$$h(\infty) = \lim_{u\to\infty} h(u)$$

exists and that

$$h'(u)h'(h(u)) < 1 \text{ for } u > A^*.$$

Then the solution $\{A_n\}$ of (5.47) with $A_0 = h(\infty)$ tends to A^* as $n \to \infty$.

Theorem 5.2.2 *Assume that $\gamma(t) \equiv 1$ and*

$$\int_0^{mT} b(t)e^{-x^*(t)} \, dt \leq 1. \tag{5.48}$$

Then every solution of (5.45) tends to $x^*(t)$ as $t \to \infty$, that is,

$$\lim_{t \to \infty} \left[x(t) - x^*(t) \right] = 0.$$

Proof Since the transformation $y(t) = x(t) - x^*(t)$ transforms (5.45) in to the equation

$$y'(t) = -a(t)y(t) + b(t)e^{-x^*(t)} \left(e^{-y(t-mT)} - 1 \right), \tag{5.49}$$

then it suffices to show that every solution of (5.49) tends to zero as $t \to \infty$.

First, we show that every nonoscillatory solution of (5.49) tends to zero. We assume that $y(t)$ is eventually positive; the proof for the case $y(t)$ is eventually negative is similar and is omitted. Since $y(t) > 0$, from (5.49) we see that there is a $T_1 > 0$ such that $y'(t) < 0$ for $t > T_1$, and so there exists a constant $l \geq 0$ such that $\lim_{t \to \infty} y(t) = l$, and

$$y'(t) \leq -la(t) + b(t)e^{-x^*(t)} \left(e^{-l} - 1 \right)$$

for $t \geq T_1 + mT$. Hence, it follows that

$$l - y(T_1 + mT) \leq -l \int_{T_1+mT}^{\infty} a(t) \, dt,$$

which, clearly, implies that $l = 0$.

Next, assume that $y(t)$ oscillates. We claim that for any $n \geq 0$, there exists a constant $T(n) \geq 0$ such that

$$y_{2n} \leq y(t) \leq y_{2n+1} \text{ for } t \geq T(n), \tag{5.50}$$

where $\{y_n\}$ is defined by

$$y_{n+1} = \left(\int_0^{mT} b(t)e^{-x^*(t)} \, dt \right) (e^{-y_n} - 1)$$

with

$$y_0 = - \int_0^{mT} b(t)e^{-x^*(t)} \, dt.$$

Since $y(t)$ is oscillatory, there is an increasing sequence $\{t_n\}$ such that

$$t_{n+1} - t_n \geq 2mT + 1 \text{ and } y(t_n) = 0 \text{ for } n = 0, 1, \ldots.$$

Let $s_n \in (t_n, t_{n+1})$ be the point where $y(t)$ obtains its local maxima or local minima in (t_n, t_{n+1}). Then $y'(s_n) = 0$ and it follows from (5.49) that

$$-a(s_n)y(s_n - mT) < 0, \quad n = 0, 1, \ldots.$$

Hence, there is a $\xi \in (s_n - mT, s_n)$ such that

$$y(\xi_n) = 0, \quad n = 0, 1, \ldots. \tag{5.51}$$

From Eq. (5.49), we see that

$$\frac{d}{dt} \left[y(t)e^{\int_0^t a(s) \, ds} \right] = b(t)e^{\int_0^t a(s) \, ds} e^{-x^*(t)} \left(e^{-y(t-mT)} - 1 \right),$$

and integrating from ξ_n to s_n and using (5.51), we obtain

$$y(s_n)e^{\int_0^{s_n} a(s) \, ds} \geq -e^{\int_0^{s_n} a(s) \, ds} \int_0^{mT} b(t)e^{-x^*(t)} \, dt. \tag{5.52}$$

Thus,

$$y(s_n) \geq - \int_0^{mT} b(t)e^{-x^*(t)} \, dt = y_0 \text{ for } n = 0, 1, \ldots. \tag{5.53}$$

Since e^{-u} is decreasing, substituting (5.53) into (5.52), we obtain

$$y(s_n)e^{\int_0^{s_n} a(s) \, ds} \leq e^{\int_0^{s_n} a(s) \, ds} \left(\int_0^{mT} b(t)e^{-x^*(t)} \, dt \right) (e^{-y_0} - 1),$$

and so

$$y(s_n) \le \left(\int_0^{mT} b(t)e^{-x^*(t)}\,dt \right) \left(e^{-y_0} - 1 \right) = y_1 \text{ for } n = 1, 2, \ldots .$$

Hence,

$$y(t) \le y_1 \text{ for } t \ge t_1$$

and so

$$y_0 \le y(t) \le y_1 \text{ for } t \ge t_1.$$

Now, assume that

$$y_{2k} \le y(t) \le y_{2k+1} \text{ for } t \ge t_{2k+1}. \tag{5.54}$$

Substituting (5.54) into (5.52) gives

$$y(s_n)e^{\int_0^{s_n} a(s)\,ds} \ge e^{\int_0^{s_n} a(s)\,ds} \left(\int_0^{mT} b(t)e^{-x^*(t)}\,dt \right) \left(e^{-y_{2k+1}} - 1 \right),$$

and so

$$y(s_n) \ge \left(\int_0^{mT} b(t)e^{-x^*(t)}\,dt \right) \left(e^{-y_{2k+1}} - 1 \right)$$

$$= y_{2(k+1)}, \quad n = 2(k+1), 2(k+1)+1, \ldots .$$

Hence,

$$y(t) \ge y_{2(k+1)} \text{ for } t \ge t_{2(k+1)}. \tag{5.55}$$

Using (5.55) in (5.52), we obtain

$$y(s_n)e^{\int_0^{s_n} a(s)\,ds} \le e^{\int_0^{s_n} a(s)\,ds} \left(\int_0^{mT} b(t)e^{-x^*(t)}\,dt \right) \left(e^{-y_{2(k+1)}} - 1 \right);$$

and so

$$y(s_n) \le \left(\int_0^{mT} b(t)e^{-x^*(t)}\,dt \right) \left(e^{-y_{2(k+1)}} - 1 \right)$$

$$= y_{2(k+1)+1}, \quad n = 2(k+1)+1, 2(k+2), \ldots .$$

Hence,

$$y_{2(k+1)} \le y(t) \le y_{2(k+1)+1} \text{ for } t \ge t_{1k+1}.$$

Therefore, by induction, we see that (5.50) holds.

Now, we claim that

$$\lim_{n \to \infty} y_n = 0.$$

To this end, let

$$h(u) = \left(\int_0^{mT} b(t) e^{-x^*(t)} \, dt \right) \left(e^{-u} - 1 \right).$$

Then,

$$h'(u) = - \left(\int_0^{mT} b(t) e^{-x^*(t)} \, dt \right) e^{-u},$$

and so

$$h'(u) h'(h(u)) = \left(\int_0^{mT} b(t) e^{-x^*(t)} \, dt \right)^2 e^{-(u+h(u))}. \tag{5.56}$$

Observe that

$$u + h(u) = u + \left(\int_0^{mT} b(t) e^{-x^*(t)} \, dt \right) \left(e^{-u} - 1 \right),$$

which, in view of (5.48) and the fact that $e^{-u} \ge 1 - u$, implies that

$$u + h(u) \ge u + \left(e^{-u} - 1 \right) > 0 \text{ for } u > 0.$$

Hence, it follows from (5.56) that

$$h'(u) h'(h(u)) < \left(\int_0^{mT} b(t) e^{-x^*(t)} \, dt \right)^2 \le 1 \text{ for } u > 0,$$

and so by Lemma 5.2.1,

$$\lim_{n \to \infty} y_n = 0.$$

Then, in view of (5.50), we see that

$$\lim_{t \to \infty} y(t) = 0.$$

This completes the proof of the theorem.   □

Theorem 5.2.3 *Let* $\gamma(t) \equiv 1$. *Then every solution of* (5.45) *oscillates about* $x^*(t)$ *if and only if*

$$e^{\int_0^{mT} a(t)\,dt} \int_0^{mT} b(t)e^{-x^*(t)}\,dt > 1/e.$$

Setting $x(t) = \gamma y(t)$, Eq. (5.42) becomes

$$x'(t) = -ax(t) + b\gamma e^{-x(t-\tau)}, \ t \geq 0. \tag{5.57}$$

Equation (5.42) has a unique positive equilibrium y^* and so $x^* = \gamma y^*$ is the unique equilibrium of (5.57). Clearly, every solution of (5.42) oscillates about y^* if and only if every solution of (5.57) oscillates about x^*; every solution of (5.42) tends to y^* if and only if every solution of (5.57) tends to x^*. Hence, by employing Theorem 5.2.3, we see that every solution of (5.42) oscillates about y^* if and only if

$$e^{a\tau}b\gamma\tau e^{-\gamma y^*} > 1/e. \tag{5.58}$$

Since y^* is an equilibrium point of (5.42),

$$ay^* = be^{-\gamma y^*}.$$

Hence, (5.58) is equivalent to

$$a\gamma\tau y^* e^{a\tau} > 1/e. \tag{5.59}$$

The condition (5.59) appears as a necessary and sufficient condition in [11] for every positive solution of (5.42) to oscillate about y^*.

By employing Theorem 5.2.2, we see that every solution of Eq. (5.45) tends to y^* as t tends to ∞ provided

$$b\tau\gamma e^{-\gamma y^*} \leq 1,$$

that is,

$$a\gamma\tau y^* \leq 1. \tag{5.60}$$

Kulenovic et al. [12] provided a different sufficient condition, namely

$$\exp\left(\gamma y^* \left(1 - e^{-a\tau}\right)\right) < 2,$$

by which every positive solution of (5.42) tends to y^* as $t \to \infty$.

Theorem 5.2.4 *Let $\gamma(t) = \gamma > 0$. Suppose that either*

$$\liminf_{t\to\infty} \int_{t-mT}^{t} \exp\left(\int_{s-mT}^{s} a(\theta)\,d\theta\right) \gamma b(s) e^{-\gamma x^*(s)}\,ds > 1/e$$

or

$$\limsup_{t\to\infty} \int_{t-mT}^{t} \exp\left(\int_{s-mT}^{s} a(\theta)\,d\theta\right) \gamma b(s) e^{-\gamma x^*(s)}\,ds > 1.$$

Then every solution of (5.45) *oscillates about* $x^*(t)$.

Theorem 5.2.5 *Let $\gamma(t) \equiv \gamma > 0$ and*

$$\lim_{t\to\infty} \int_{t-mT}^{t} \gamma b(s) \exp\left(\int_{s-mT}^{s} a(\theta)\,d\theta\right) ds < \pi/2.$$

Then every solution $x(t)$ of (5.45) *satisfies*

$$\lim_{t\to\infty} \left[x(t) - x^*(t)\right] = 0.$$

Li and Wang [15] used Mawhin's continuation theorem and proved that (5.43) has at least one positive T-periodic solution. The following theorem provides a sufficient condition under which a periodic solution of (5.43) is a global attractor.

Theorem 5.2.6 *Assume that $\tau'(t) \le 1$ and*

$$\gamma(t)b(t) \le a(t).$$

Then, any positive T-periodic solution of (5.43) *is a global attractor.*

Let X be the Banach space $\{x \in C(R, R) : x(t + T) = x(t), \ t \in R\}$ with the sup norm $\|x\| = \sup_{0 \le t \le T} |x(t)|$, and define the cone K by

$$K = \{x \in X : x(t) \ge 0, \ t \in [0, T]\}.$$

Let the operator A be defined by

$$(Ax)(t) = \int_{t}^{t+T} G(t, s)b(s) e^{-\gamma(s)x(s-\tau(s))}\,ds,$$

where $G(t, s)$ is the Green's kernel given by

$$G(t, s) = \frac{e^{\int_t^s a(\theta)\, d\theta}}{e^{\int_0^T a(\theta)\, d\theta} - 1}.$$

Set

$$\beta = \frac{e^{\int_0^T a(\theta)\, d\theta}}{e^{\int_0^T a(\theta)\, d\theta} - 1}, \quad \overline{\gamma} = \max_{0 \le t \le T} \gamma(t), \text{ and } b = \int_0^T b(t)\, dt.$$

Theorem 5.2.7 *Let* $\beta b \overline{\gamma} \le 1$. *Then, Eq. (5.43) has a unique positive* T*-periodic solution* $x^*(t)$. *Moreover,*

$$\lim_{n \to \infty} \|x_n - x^*\| = 0,$$

where $x_n = Ax_{n-1}$ $(n = 1, 2, \ldots)$ *for any initial* $x_0 \in K$,

Theorem 5.2.8 *Assume that* $\beta b \overline{\gamma} \le 1$. *Let* $x^*(t)$ *be the unique positive* T*-periodic solution of* (5.43). *Then every solution* $x(t)$ *of* (5.43) *satisfies*

$$\lim_{t \to \infty} [x(t) - x^*(t)] = 0.$$

5.3 Global Attractivity of Periodic Solutions of a Red Blood Cell Production Model

This section is concerned with the global attractivity of a positive periodic solution of the red blood cell production model (5.2). This model was first proposed by Mackey and Glass [19] to describe some physiological control systems with constant coefficients and constant delay.

There has been some remarkable work (see [27] and the references cited therein) on determining necessary and sufficient conditions for the existence of periodic solutions of Eq. (5.2).

Wang and Li [25] investigated the existence and global attractivity of the unique positive periodic solution $\overline{x}(t)$ of the equation

$$x'(t) = -a(t)x(t) + \frac{b(t)}{1 + x^n(t - \tau(t))}, \quad n > 1. \tag{5.61}$$

In [17], the authors consider a generalized form of (5.61), namely,

$$x'(t) = -a(t)x(t) + \sum_{i=1}^{m} \frac{b_i(t)}{1 + x^n(t - \tau_i(t))}, \quad n > 0 \qquad (5.62)$$

and established existence and global attractivity of a unique positive periodic solution.

Rost [22] and Saker [24] studied the oscillation and global attractivity of solutions of the constant coefficient model

$$x'(t) = -ax(t) + b\frac{x^m(t - \tau)}{1 + x^n(t - \tau)}. \qquad (5.63)$$

Now, we will provide some results on the global attractivity of positive periodic solution of (5.2) by constructing a suitable Lyapunov function.

Consider the solutions of (5.2) with the initial function

$$x(t) = \phi(t), \quad t \in [-\tau, 0], \quad \phi \in C([-\tau, 0], [0, \infty)), \quad \phi(0) > 0. \qquad (5.64)$$

Throughout the section, we assume that $n > 1$.

As the global attractivity of positive periodic solution of (5.2) is our objective, we use the following lemma for the existence of one positive T-periodic solution of (5.2) discussed in [27, 28], and we show that this periodic solution is a global attractor to all other positive solutions.

Lemma 5.3.1 ([27]) *Assume that $a(t)$, $b(t) \in C(R, R_+)$ are T-periodic functions, τ and n are positive constants, and $b(t) > a(t)$ for $t \in [0, T]$. Then the initial value problem (5.2), (5.64) has at least one positive T-periodic solution.*

Before proving the global attractivity of the positive periodic solution of (5.2), we give a lemma for the positiveness and boundedness of the solutions.

Lemma 5.3.2 *Every solution of (5.2), (5.64) is positive and bounded.*

Proof Suppose that $x(t)$ is a solution of (5.2) with initial function (5.64). If $x(t)$ is not eventually positive, then from the initial condition (5.64), it follows that there exists a $t_1 > 0$ such that $x(t) > 0$ for $t \in [0, t_1)$, $x(t_1) = 0$ and $x'(t_1) < 0$. Hence from (5.2) and (5.64), we have

$$x''(t_1) = \frac{b(t_1)x(t_1 - \tau)}{1 + x^n(t_1 - \tau)} > 0,$$

which is a contradiction. Hence, $x(t) > 0$ for all $t \geq t_0$.

Let $F(x) = \frac{x}{1+x^n}$. It is easy to see that F attains its maximum $F_{\max} = \frac{\mu}{1+\mu^n}$ where $\mu = \left(\frac{1}{n-1}\right)^{1/n}$. From (5.2), we have

$$x(t) = \phi(0)e^{-\int\limits_0^t a(\theta)\,d\theta} + \int\limits_0^t b(s)\frac{x(s-\tau)}{1+x^n(s-\tau)}e^{-\int\limits_s^t a(\theta)\,d\theta}\,ds$$

$$\leq \phi(0)e^{-a_*t} + b^* F_{\max} \int\limits_0^t e^{-a_*(t-s)}\,ds$$

$$\leq \phi(0)e^{-a_*t} + \frac{b^*}{a_*}F_{\max}[1 - e^{-a_*t}].$$

Hence,

$$\limsup_{t\to\infty} x(t) \leq \frac{b^*}{a_*}F_{\max} = \frac{b^*}{a_*n}(n-1)^{\frac{n-1}{n}}.$$

Therefore, $x(t)$ is bounded and this proves the lemma.   □

Remark 5.3.1 Lemma 5.3.2 yields that every solution of (5.2) is bounded by $\frac{b^*}{a_*n}$ $(n-1)^{\frac{n-1}{n}}$.

Set

$$M = \frac{b^*}{a_*n}(n-1)^{\frac{n-1}{n}}. \tag{5.65}$$

In the following, by means of a Lyapunov function, we obtain a sufficient condition for the global attractivity of a positive periodic solution $\overline{x}(t)$ of (5.2).

Theorem 5.3.1 *If*

$$a_*(1+M^n)^2 + b^* > (n-1)b^* M^n \tag{5.66}$$

and

$$(a_*n)^n b^{*^n}(n-1)^n < (a_*n)^n(n+1), \tag{5.67}$$

then every solution of (5.2) tends to $\overline{x}(t)$ as t tends to ∞, that is,

$$\lim_{t\to\infty}[x(t) - \overline{x}(t)] = 0,$$

where M is given in (5.65).

Proof Let $x(t)$ be a solution of (5.2). Suppose that $z(t) = x(t) - \overline{x}(t)$ for $t \geq t_0 > (\frac{1}{n-1})^{\frac{1}{n}}$. Then,

$$z'(t) = -a(t)[x(t) - \overline{x}(t)] + b(t)\left[\frac{x(t-\tau)}{1+x^n(t-\tau)} - \frac{\overline{x}(t-\tau)}{1+\overline{x}^n(t-\tau)}\right], \quad t \geq t_0.$$

To complete the proof of the theorem, it suffices to show that $z(t) \to 0$ as $t \to \infty$. Define a Lyapunov function by

$$V(t) = |z(t)| + \int_{t-\tau}^{t} b(s+\tau) \left| \frac{x(s)}{1+x^n(s)} - \frac{\overline{x}(s)}{1+\overline{x}^n(s)} \right| ds$$

$$= |z(t)| + \int_{t-\tau}^{t} b(s+\tau) \left| \frac{z(s)+\overline{x}(s)}{1+[z(s)+\overline{x}(s)]^n} - \frac{\overline{x}(s)}{1+\overline{x}^n(s)} \right| ds.$$

Then,

$$\frac{dV}{dt} = sgn[z(t)]\left[-a(t)z(t) \right.$$

$$+ b(t) \left\{ \frac{z(t-\tau)+\overline{x}(t-\tau)}{1+[z(t-\tau)+\overline{x}(t-\tau)]^n} - \frac{\overline{x}(t-\tau)}{1+\overline{x}^n(t-\tau)} \right\} \right] \qquad (5.68)$$

$$+ b(t+\tau) \left| \frac{z(t)+\overline{x}(t)}{1+[z(t)+\overline{x}(t)]^n} - \frac{\overline{x}(t)}{1+\overline{x}^n(t)} \right|$$

$$- b(t) \left| \frac{z(t-\tau)+\overline{x}(t-\tau)}{1+[z(t-\tau)+\overline{x}(t-\tau)]^n} - \frac{\overline{x}(t-\tau)}{1+\overline{x}^n(t-\tau)} \right|$$

$$\leq -a(t)|z(t)| + b(t+\tau) \left| \frac{z(t)+\overline{x}(t)}{1+[z(t)+\overline{x}(t)]^n} - \frac{\overline{x}(t)}{1+\overline{x}^n(t)} \right|$$

$$\leq -a(t)|z(t)| + b(t+\tau)|z(t)| \frac{(n-1)\theta^n-1}{(1+\theta^n)^2}, \qquad (5.69)$$

where θ lies between $z(t)+\overline{x}(t)$ and $\overline{x}(t)$. Setting $G(\theta) = \frac{(n-1)\theta^n-1}{(1+\theta^n)^2}$, we see that $G(0) = -1$, $G\left((\frac{1}{n-1})^{\frac{1}{n}}\right) = 0$, and $G(\theta)$ is positive and increasing for $\theta > (\frac{1}{n-1})^{\frac{1}{n}}$. Furthermore, condition (5.67) implies that

$$\left(\frac{1}{n-1}\right)^{\frac{1}{n}} < M < \left(\frac{n+1}{n-1}\right)^{\frac{1}{n}}.$$

Hence, we have

$$\frac{dV}{dt} \leq -\left[a_* - b^* \frac{[(n-1)M^n-1]}{(1+M^n)^2} \right] |z(t)|$$

$$\leq -\frac{1}{(1+M^n)^2}[a_*(1+M^n)^2 + b^* - b^*(n-1)M^n]|z(t)|.$$

Integrating the above inequality from t_0 to t, we have

$$V(t) + \frac{1}{(1+M^n)^2}[a_*(1+M^n)^2 + b^* - b^*(n-1)M^n] \int_{t_0}^{t} |z(s)| ds \leq V(t_0).$$

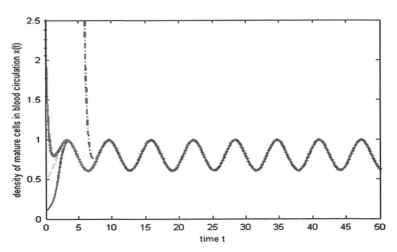

Fig. 5.5 Computer simulation of Example 5.3.1

Thus, $|z(t)| \in L([t_0, \infty))$ as $t \to \infty$, i.e, $\int_{t_0}^{\infty} |z(s)|\, ds < \infty$. By [5, Lemma 1.2.2] of Gopalsamy, it follows that $\lim\limits_{t \to \infty} z(t) = 0$. This completes the proof of the theorem.

Next, we give an example to illustrate Theorem 5.3.1.

Example 5.3.1 Consider the equation

$$x'(t) = -\left(2 + \frac{\cos t}{2}\right) x(t) + \left(3 + \frac{\sin t}{4}\right) \frac{x(t-\tau)}{1 + x^3(t-\tau)}, \quad t \geq 0. \qquad (5.70)$$

Here $a(t) = 2 + \frac{\cos t}{2}$, $b(t) = 3 + \frac{\sin t}{4}$, $n = 3$, $a_* = 1.5$, $a^* = 2.5$, $b_* = 2.75$, $b^* = 3.25$. Clearly $b_* > a^*$. Also, $M = \frac{b^*}{a_* n}(n-1)^{\frac{n-1}{n}} = 1.146456315$ and $a_*(1 + M^n)^2 + b^* = 12.677 > 9.795 = b^*(n-1)M^n$, and $b^{*^n}(n-1)^n = 274.625 < 364.5 = (a_* n)^n (n+1)$. Hence, by Lemma 5.3.1, Eq. (5.70) has a positive periodic solution $\overline{x}(t)$ and this solution is globally attractive to all other positive solutions of (5.70) by Theorem 5.3.1 (Fig. 5.5).

Remark 5.3.2 Using the property that $\frac{x^{n-1}}{1+x^n} \leq 1$ for $x \geq 0$ in (5.69), we can obtain the following theorem.

Theorem 5.3.2 *If*

$$a_*(1 + M^n)^2 > b^*[(n-1)M(1 + M^n)^2 - 1], \qquad (5.71)$$

where M is given in (5.65), then every solution of (5.2) tends to $\overline{x}(t)$ as t tends to ∞, that is,

$$\lim_{t \to \infty} [x(t) - \overline{x}(t)] = 0.$$

Remark 5.3.3 Qian [21] proved that if

$$b \max \left\{ 1, \frac{(n-1)^2}{4n} \right\} \tau < 1, \tag{5.72}$$

then any positive periodic solution of (5.2) is a global attractor of any other periodic solution of (5.2) (see [21, Example 2]). Now, consider the equation

$$x'(t) = -2.5x(t) + 3.25 \frac{x(t-1)}{1 + x^3(t-1)}, \quad t \geq 0. \tag{5.73}$$

Clearly, the conditions (5.66) and (5.67) are satisfied and hence Theorem 5.3.1 can be applied to (5.73). On the other hand, (5.72) does not hold and hence the result due to Qian [21] cannot be applied to this example.

Consider (5.2) with $b(t) = b = \sqrt{2}$, $a(t) = a = 1$, $\tau = 1$, and $n = 2$. Theorem 5.3.2 can be applied to this example but condition (5.72) does not hold.

5.4 Global Attractivity of Periodic Solutions of Nicholson's Blowflies Model

This section is concerned with the dynamics of the Eq. (5.3) with constant delay, where $a, b \in C(R_+, R_+)$ are periodic functions with period $T > 0$ and $\gamma, \tau \in (0, \infty)$.

Gurney et al. [8] used the Nicholson's blowflies model (5.3) with constant coefficients to describe the population of Australian sheep blowflies that agrees well with the experimental data of Nicholson [20].

In [13] Kulenovic et al. proved that if $b > a$, then every positive solution $x(t)$ of (5.3) tends to $\bar{x}$ as $t \to \infty$ provided that

$$(e^{a\tau} - 1) \left(\frac{b}{a} - 1 \right) < 1,$$

holds, where $\bar{x}$ is an equilibrium point of (5.3).

Existence of a periodic solution of (5.3) with constant coefficients is studied in Chap. 2 while dealing with the first order functional differential equation (2.1), where $a(t)$ and $b(t)$ are positive T-periodic functions, $\tau > 0$ is a constant and $f \in C(R, R_+)$.

Saker and Agarwal [23] obtained sufficient conditions for the oscillation and global attractivity of the periodic Nicholson's blowflies model

$$x'(t) = -a(t)x(t) + b(t)x(t - mT)e^{-\gamma x(t-mT)}. \tag{5.74}$$

They established the following sufficient condition for the global attractivity of a positive equilibrium point of (5.74).

Theorem 5.4.1 *Assume that $a(t)$ and $b(t)$ are T-periodic functions and $\gamma \bar{x}_* > 1$. Assume that*

$$\max_{0 \le t \le T} \frac{b(t)}{a(t)} e^{-\gamma \bar{x}(t)} \le 1 \tag{5.75}$$

and

$$\limsup_{t \to \infty} \int_{t-mT}^{t} b(s) \exp\left(\int_{t}^{s} a(u)\, du \right) ds = \gamma < 1. \tag{5.76}$$

Then $\lim_{t \to \infty}[x(t) - \bar{x}(t)] = 0$ for any positive solution $x(t)$ of (5.74), where $\bar{x}(t)$ is positive periodic solution and $\bar{x}_* = \min_{0 \le t \le T} \bar{x}(t)$.

We begin the discussion of some more results of this section with the following lemma.

Lemma 5.4.1 *Every solution of (5.3), (5.64) is positive and bounded.*

Proof Suppose that $x(t)$ is solution of (5.3) with initial function (5.64). Then from (5.3), we have

$$x(t) = \phi(0) e^{\int_0^t \left[-a(s) + b(s) \frac{x(s-\tau)}{x(s)} e^{-\gamma x(s-\tau)} \right] ds}$$

which is positive for $t > 0$.

Next, we show that $x(t)$ is bounded. Let $F(\tilde{x}) = \tilde{x} e^{-\gamma \tilde{x}}$. It can easily be shown that F attains its maximum $F_{\max} = \frac{1}{\gamma e}$ at $\tilde{x} = \frac{1}{\gamma}$. Equation (5.3) can be written as

$$\left(x(t) e^{\int_0^t a(\theta)\, d\theta} \right)' = b(t) x(t-\tau) e^{-\gamma x(t-\tau)} e^{\int_0^t a(\theta)\, d\theta}.$$

Integrating the above equation from 0 to t, we obtain

$$x(t) = \phi(0) e^{-\int_0^t a(\theta)\, d\theta} + \int_0^t b(s) x(s-\tau) e^{-\gamma x(s-\tau)} e^{-\int_s^t a(\theta)\, d\theta}\, ds$$

$$\le \phi(0) e^{-a_* t} + b^* F_{\max} \int_0^t e^{-a_*(t-s)}\, ds$$

$$\le \phi(0) e^{-a_* t} + \frac{b^*}{a_*} F_{\max}[1 - e^{-a_* t}].$$

Hence, $\limsup_{t \to \infty} x(t) \le \frac{b^*}{a_* \gamma e}$. This completes the proof of the lemma. $\qquad \square$

Remark 5.4.1 It follows from Lemma 5.4.1 that, if $x(t)$ is a solution of (5.3), then

$$\limsup_{t \to \infty} |x(t)| \le \frac{b^*}{a_* \gamma e} =: N.$$

Lemma 5.4.2 ([28]) *Assume that $a(t)$, $b(t) \in C(R, R_+)$ are T-periodic functions, τ is a positive constant, and $b(t) > a(t)$ for $t \in [0, T]$. Then (5.3), (5.64) has at least one positive T-periodic solution.*

Let this solution be $\bar{x}(t)$. We use the boundedness property of positive solutions of (5.3) from Remark 5.4.1 in the following theorem.

Theorem 5.4.2 *Assume that $\frac{a_*}{b^*} e^{\frac{b^*}{a_* e}} + 1 > \frac{b^*}{a_* e}$ and $b^* < 2a_* e$. Then every solution of (5.3) tends to $\bar{x}(t)$ as t tends to ∞.*

Proof Let $x(t)$ be an arbitrary solution of (5.3). Suppose $z(t) = x(t) - \bar{x}(t)$ for some $t \ge t_0 > \frac{1}{\gamma}$; then

$$z'(t) = -a(t)z(t) + b(t)[x(t-\tau)e^{-\gamma x(t-\tau)} - \bar{x}(t-\tau)e^{-\gamma \bar{x}(t-\tau)}], \quad t \ge t_0.$$

To complete the proof of the theorem, it suffices to show that $z(t) \to 0$ as $t \to \infty$.

Define the Lyapunov function $V(t)$ by

$$V(t) = |z(t)| + \int_{t-\tau}^{t} b(s+\tau)|x(s)e^{-\gamma x(s)} - \bar{x}(s)e^{-\gamma \bar{x}(s)}| \, ds$$

$$= |z(t)| + \int_{t-\tau}^{t} b(s+\tau)|(z(s)+\bar{x}(s))e^{-\gamma(z(s)+\bar{x}(s))} - \bar{x}(s)e^{-\gamma \bar{x}(s)}| \, ds.$$

Then,

$$\frac{dV}{dt} = sgn[z(t)][-a(t)z(t) + b(t)\{(z(t-\tau)+\bar{x}(t-\tau))e^{-\gamma(z(t-\tau)+\bar{x}(t-\tau))}$$

$$-\bar{x}(t-\tau)e^{-\gamma \bar{x}(t-\tau)}\}] + b(t+\tau)\left|(z(t)+\bar{x}(t))e^{-\gamma(z(t)+\bar{x}(t))} - \bar{x}(t)e^{-\gamma \bar{x}(t)}\right|$$

$$- b(t)\left|(z(t-\tau)+\bar{x}(t-\tau))e^{-\gamma(z(t-\tau)+\bar{x}(t-\tau))} - \bar{x}(t-\tau)e^{-\gamma \bar{x}(t-\tau)}\right|$$

$$\le - a(t)|z(t)| + b(t+\tau)|z(t)||F'(\theta),$$

where $F(\theta) = \theta e^{-\gamma \theta}$ and θ lies between $z(t) + \bar{x}(t)$ and $\bar{x}(t)$. It is clear that $F'(\theta) = (1 - \gamma\theta)e^{-\gamma\theta} < 0$ for $\theta > \frac{1}{\gamma}$. Hence, setting

$$G_1(\theta) = -F'(\theta) = (\gamma\theta - 1)e^{-\gamma\theta} > 0,$$

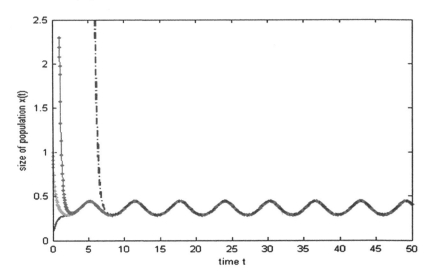

Fig. 5.6 Computer simulation of Example 5.4.1

it follows from the fact that $b^* < 2a_*e$ that $G_1(\theta)$ is increasing for $0 < \theta < \frac{2}{\gamma}$. Then,

$$\frac{dV}{dt} \leq [-a(t) + b(t+\tau)G_1(\theta)]|z(t)|$$
$$\leq [-a_* + b^*(\gamma\theta - 1)e^{-\gamma\theta}]|z(t)|$$
$$\leq -[a_* - b^*(\gamma N - 1)e^{-\gamma N}]|z(t)|.$$

Integrating from t_0 to t, we obtain

$$V(t) + [a_* - b^*(\gamma N - 1)e^{-\gamma N}] \int_{t_0}^{t} |z(s)|\, ds \leq V(t_0).$$

This implies that $|z(t)| \in L([t_0, \infty))$ as $t \to \infty$, that is, $\int_{t_0}^{\infty} |z(s)|\, ds < \infty$. Then by Gopalsamy [5, Lemma 1.2.2], $\lim_{t\to\infty} z(t) = 0$. This completes the proof of the theorem.   □

Example 5.4.1 Consider the differential equation

$$x'(t) = -(3 + \sin t)x(t) + (6 + \sin t)x(t - \tau)e^{-2x(t-\tau)}. \tag{5.77}$$

Here $a(t) = 3 + \sin t$, $b(t) = 6 + \sin t$, $a_* = 2$, $a^* = 4$, $b^* = 7$, and $b_* = 5$. Clearly, $b_* > a^*$, $\frac{a_*}{b^*}e^{\frac{b^*}{a_*e}} + 1 = 2.035428 > 1.2875 = \frac{b^*}{a_*e}$, and $b^* = 7 < 2a_*e$. This shows that Theorem 5.4.2 can be applied to this example (Fig. 5.6).

Next, we give an example where Theorem 5.4.1 fails whereas Theorem 5.4.2 can be applied. We see that Eq. (5.74) is equivalent to the Eq. (5.3) for $mT = \tau$.

Example 5.4.2 Consider the equation

$$x'(t) = -\left(\frac{3}{2} + \sin t\right) x(t) + (4 + \sin t)x(t-1)e^{-\gamma x(t-1)}. \qquad (5.78)$$

Here $a(t) = \frac{3}{2} + \sin t$, $b(t) = 4 + \sin t$, $a_* = \frac{1}{2}$, $a^* = \frac{5}{2}$, $b_* = 3$, $b^* = 5$, and $\tau = 1$. Then a simple calculation shows that $\limsup\limits_{t\to\infty} \int_{t-1}^{t}(4 + \sin v)\exp(\int_{t}^{v}(3/2 + \sin u)\,du)\,dv > 2.360186$, which means that (5.76) fails to hold. On the other hand, $b^* > a_*$ and $\frac{a_*}{b^*}e^{\frac{b^*}{a_* e}} + 1 = 4.950671 > 3.67647 = \frac{b^*}{a_* e}$, hold. Hence, (5.78) has a positive 2π-periodic solution and this solution is globally attractive to all other positive solutions of (5.78) by Lemma 5.4.1 and Theorem 5.4.2.

Appendix

Program for Example 5.1.1

```
function dydt = sh(t,y)
dydt =...
    -(1+sin(t).^2/10).*y+(y.*(1.5+cos(t).^2./20))./(1+(y./(2+cos(t).
    ^2)).^3);

clear all
clc

t0 = 0;
y0 = 0.1;

[t,y] = ode45(@sh,[t0 100],y0);
plot(t,y,'LineWidth',2);

t0 = 0;
y0 = 0.5;

hold on;
[t,y] = ode45(@sh,[t0 100],y0);
plot(t,y,'--g','LineWidth',2);

t0 = 0;
y0 = 4.5;

hold on;
[t,y] = ode45(@sh,[t0 100],y0);
plot(t,y,'-r','LineWidth',2);

t0 = 10;
y0 = 4.5;
```

```
hold on;
[t,y] = ode45(@sh,[t0 100],y0);
plot(t,y,'-b','LineWidth',2);

xlabel('time t');
ylabel('population x(t)');
```

Program for Example 5.1.2

```
function dydt = fish2(t,y)
dydt = ...
    -(1.2+sin(t).^2/50).*y+(y.*(1.3+cos(t).^2./100))./(1+(y./
    (0.5+sin(t).^2./1000)).^6);

clear all
clc

t0 = 0;
y0 = 0.1;

[t,y] = ode45(@fish2,[t0 40],y0);
plot(t,y,'LineWidth',2);

t0 = 0;
y0 = 0.5;

hold on;
[t,y] = ode45(@fish2,[t0 40],y0);
plot(t,y,'.-g','LineWidth',2);

t0 = 0;
y0 = 4.5;

hold on;
[t,y] = ode45(@fish2,[t0 40],y0);
plot(t,y,'-r','LineWidth',2);

t0 = 5;
y0 = 4.5;

hold on;
[t,y] = ode45(@fish2,[t0 40],y0);
plot(t,y,'-b','LineWidth',2);

xlabel('time t');
ylabel('population x(t)');
```

Program for Example 5.1.3

```
function dydt = fish3(t,y)
dydt =
    -(1.5+sin(t).^2/10).*y+(y.*(1.7+cos(t).^2./20))./(1+(y./(4+sin(t).^2./100)).^6);

clear all
clc

t0 = 0;
y0 = 0.1;

[t,y] = ode45(@fish3,[t0 80],y0);
plot(t,y,'LineWidth',2);

t0 = 0;
```

```
y0 = 0.5;

hold on;
[t,y] = ode45(@fish3,[t0 80],y0);
plot(t,y,'-.g','LineWidth',2);

t0 = 0;
y0 = 4.5;

hold on;
[t,y] = ode45(@fish3,[t0 80],y0);
plot(t,y,'-r','LineWidth',2);

t0 = 5;
y0 = 4.5;

hold on;
[t,y] = ode45(@fish3,[t0 80],y0);
plot(t,y,'-b','LineWidth',2);

xlabel('time t');
ylabel('population x(t)');
```

Program for Example 5.1.4

```
function dydt = fish4(t,y)
dydt = -y+(2.*y)./(1+y.^2);

clear all
clc

t0 = 0;
y0 = 0.1;

[t,y] = ode45(@fish4,[t0 30],y0);
plot(t,y,'LineWidth',2);

t0 = 0;
y0 = 0.5;

hold on;
[t,y] = ode45(@fish4,[t0 30],y0);
plot(t,y,'--g','LineWidth',2);

t0 = 0;
y0 = 4.5;

hold on;
[t,y] = ode45(@fish4,[t0 30],y0);
plot(t,y,'.-r','LineWidth',2);

t0 = 5;
y0 = 4.5;

hold on;
[t,y] = ode45(@fish4,[t0 30],y0);
plot(t,y,'-.b','LineWidth',2);

xlabel('time t');
ylabel('population x(t)');
```

Program for Example 5.3.1

```
function dydt = RBC(t,y)
dydt = -(2+cos(t)./2).*y+(y.*(3+sin(t)./4))./(1+y.^3);

clear all
clc

t0 = 0;
y0 = 0.1;

[t,y] = ode45(@RBC,[t0 50],y0);
plot(t,y,'LineWidth',2);

t0 = 0;
y0 = 0.5;

hold on;
[t,y] = ode45(@RBC,[t0 50],y0);
plot(t,y,'--g','LineWidth',2);

t0 = 0;
y0 = 2.5;

hold on;
[t,y] = ode45(@RBC,[t0 50],y0);
plot(t,y,'-r','LineWidth',2);

t0 = 6;
y0 = 2.5;

hold on;
[t,y] = ode45(@RBC,[t0 50],y0);
plot(t,y,'-.b','LineWidth',2);

xlabel('time t');
ylabel('density of mature cells in blood circulation x(t)');
```

Program for Example 5.4.1

```
function dydt = nich(t,y)
dydt = -(3+sin(t)).*y+y.*(6+sin(t)).*exp(-2.*y);

clear all
clc

t0 = 0;
y0 = 0.1;

[t,y] = ode45(@nich,[t0 50],y0);
plot(t,y,'LineWidth',2);

t0 = 0;
y0 = 0.5;

hold on;
[t,y] = ode45(@nich,[t0 50],y0);
plot(t,y,'--g','LineWidth',2);

t0 = 1;
y0 = 2.3;

hold on;
[t,y] = ode45(@nich,[t0 50],y0);
```

```
plot(t,y,'-r','LineWidth',2);

t0 = 6;
y0 = 2.5;

hold on;
[t,y] = ode45(@nich,[t0 50],y0);
plot(t,y,'-.b','LineWidth',2);

xlabel('time t');
ylabel('size of population x(t)');
```

References

1. Berezansky, L., Idels, L.: Population models with delay in dynamic environment. Int. J. Qual. Theory Differ. Equ. Appl. **1**, 19–27 (2007).
2. Berezansky, L., Idels, L.: Stability of a time-varying fishing model with delay. Appl. Math. Lett. **21**, 447–452 (2008)
3. Berezansky, L., Braverman, E., Idels, L.: Delay differential equations with Hill's type growth rate and linear harvesting. Comput. Math. Appl. **49**, 549–563 (2005)
4. Driver, R.D.: Ordinary and Delay Differential Equations. Springer, New York (1976)
5. Gopalsamy, K.: Stability and Oscillations in Delay Differential Equations of Population Dynamics. Kluwer, Boston (1992)
6. Graef, J.R., Padhi, S., Srivastava, S.: Dynamics for a time varying fishing model. J. Neural Parallel Sci. Comput. **18**, 109–120 (2010)
7. Graef, J.R., Qian, C., Spikes, P.W.: Oscillation and global attractivity in a periodic delay equation. Can. Math. Bull. **38**, 275–283 (1996)
8. Gurney, W.S.C., Blathe, S.P., Nishet, R.M.: Nicholson's blowflies revisited. Nature **287**, 17–21 (1980)
9. Kot, M.: Elements of Mathematical Ecology. Cambridge University Press, Cambridge (2001)
10. Kuang, Y.: Delay Differential Equations with Applications in Population Dynamics. Academic Press, New York (1993)
11. Kulenovic, M.R.S., Ladas, G.: Linearized oscillations in population dynamics. Bull. Math. Biol. **49**, 615–627 (1987)
12. Kulenovic, M.R.S., Ladas, G., Sficas, Y.G.: Global attractivity in population dynamics. Comput. Math. Appl. **18**, 925–928 (1989)
13. Kulenovic, M.R.S., Ladas, G., Sficas, Y.G.: Global attractivity in Nicholson's blowflies. Appl. Anal. **43**, 109–124 (1992)
14. Lazer, A.C.: The behavior of solutions of the differential equation $y''' + p(x)y' + q(x)y = 0$. Pac. J. Math. **17**, 435–466 (1966)
15. Li, J.W., Wang, Z.C.: Existence and global attractivity of positive periodic solutions of a survival model of red blood cells. Comput. Math. Appl. **50**, 41–47 (2005)
16. Liu, X., Takeuchi, Y.: Periodicity and global dynamics of an impulsive delay Lasota-Wazewska model. J. Math. Anal. Appl. **327**, 326–341 (2007)
17. Liu, G., Yan, J., Zhang, F.: Existence and global attractivity of unique positive periodic solution for a model of hematopoiesis. J. Math. Anal. Appl. **334**, 157–171 (2007)
18. Liu, G., Zhao, A., Yan, J.: Existence and global attractivity of unique positive periodic solution for a Lasota-Wazewska model. Nonlinear Anal. **64**, 1737–1746 (2006)
19. Mackey, M.C., Glass, L.: Oscillation and chaos in psychological control system. Science **197**, 287–289 (1977)
20. Nicholsons, A.J.: The balance of animal population. J. Animal Ecol. **2**, 132–178 (1993)

21. Qian, C.: Global attractivity in a nonlinear delay differential equation with applications. Non-linear Anal. **71**, 1893–1900 (2009)
22. Rost, G.: On the global attractivity controversy for a delay model of hematopoiesis. Appl. Math. Comput. **190**, 846–850 (2007)
23. Saker, S.H., Agarwal, S.: Oscillation and global attractivity in a periodic Nicholson's blowflies model. Math. Comput. Model. **35**, 719–731 (2002)
24. Saker, S.H.: Oscillation and global attractivity in hematopoiesis model with delay time. Appl. Math. Comput. **136**, 241–250 (2003)
25. Wang, X., Li, Z.: Dynamics for a class of general hematopoiesis model with periodic coefficients. Appl. Math. Comput. **186**, 460–468 (2007)
26. Wazewska-Czyzewska, M., Lasota, A.: Mathematical problems of the dynamics of red blood cells systems. Ann. Polish Math. Soc. Ser. III Appl. Math. **17**, 23–40 (1988)
27. Wu, X.M., Li, J.W., Zhou, H.Q.: A necessary and sufficient condition for the existence of positive periodic solutions of a model of hematopoiesis. Comput. Math. Appl. **54**, 840–849 (2007)
28. Zhang, W., Zhu, D., Bi, P.: Existence of periodic solutions of a scalar functional differential equation via a fixed point theorem. Math. Comput. Model. **46**, 718–729 (2007)

Bibliography

1. Agarwal, R.P., Berezansky, L., Braverman, E., Domoshnitsky, A.: Nonoscillation Theory of Functional Differential Equations with Applications. Springer, New York (2012)
2. Burton, T.A.: Stability and Periodic Solutions of Ordinary and Functional-Differential Equations, Mathematics in Science and Engineering, vol. 178. Academic Press, Orlando (1985)
3. Domoshnitsky, A.: Maximum principles and nonoscillation intervals for first order Volterra functional differential equations. Dyn. Contin. Discrete Impuls. Syst. Ser. A Math. Anal. **15**, 769–814 (2008)
4. Gopalsamy, K., Trofimchuk, S.I.: Almost periodic solutions of Lasota-Wazewska type delay differential equation. J. Math. Anal. Appl. **237**, 106–127 (1999)
5. Gopalsamy, K., Kulenovic, M.R.S., Ladas, G.: Environmental periodicity and time delays in a food-limited population model. J. Math. Anal. Appl. **147**, 545–555 (1990)
6. Graef, J.R., Padhi, S., Pati, S.: Periodic solutions of some models with strong Allee effects. Nonlinear Anal. Real World Appl. **13**, 569–581 (2012)
7. Graef, J.R., Padhi, S., Pati, S., Kar, P.K.: Positive solutions of differential equations with unbounded Green's kernel. Appl. Anal. Discrete Math. **6**, 159–173 (2012)
8. Graef, J.R., Padhi, S., Pati, S.: Existence and nonexistence of multiple positive periodic solutions of first order differential equations with unbounded Green's kernel. Pan. Amer. Math. J. **23**, 45–55 (2013)
9. Graef, J.R., Padhi, S., Pati, S.: Multiple positive periodic solutions of first order ordinary differential equations with unbounded Green's kernel. Comm. Appl. Anal. **17**, 319–330 (2013)
10. Gusarenko, S.A., Domoshnitskii, A.I.: Asymptotic and oscillation properties of first-order linear scalar functional-differential equations. (Russian) Differential. Uravn. **25**, 2090–2103, 2206 (1989); translation in Differ. Equ. **25**(1989), 1480–1491 (1990)
11. Kang, S.G., Zhang, G., Shi, B.: Existence of three periodic positive solutions for a class of integral equations with parameters. J. Math. Anal. Appl. **323**, 654–665 (2006)
12. Lisena, B.: Periodic solutions of logistic equations with time delay. Appl. Math. Lett. **20**, 1070–1074 (2007)
13. Liu, B.: New results on the almost periodic solutions for a model of hematopoiesis. Nonlinear Anal. Real World Appl. doi:10.1016/j.nonrwa.2013.12.003
14. Liu, X.L., Zhang, G., Cheng, S.S.: Existence of triple positive periodic solutions of a functional differential equation depending on a parameter. Abstr. Appl. Anal. **10**, 897–905 (2004)
15. Mackey, M.C., Milton, J.: Feedback delays and the origin of blood cell dynamics. Comm. Theor. Bio. **1**, 299–327 (1990)
16. Myers, R., Mackenzie, B., Bowen, K.: What is the carrying capacity for fish in the ocean? A meta-analysis of population dynamics of North Atlantic cod. Can. J. Fish. Aquat. Sci. **58**, 1464–1476 (2001)
17. Padhi, S., Pati, S., Srivastava, S.: Multiple positive periodic solutions for nonlinear first order functional difference equations. Inter. J. Dynam. Syst. Diff. Eqs. **2**, 98–114 (2009)

18. Pati, S., Graef, J.R., Padhi, S., Kar, P.K.: Periodic solutions of a single species renewable resources under periodic habitat fluctuations with harvesting and Allee effect. Comm. Appl. Nonlinear Anal. **20**, 1–16 (2013)
19. Qian, C.: Global attractivity in a variable coefficient nonlinear delay differential equation. Comm. Appl. Anal. **20**, 33–44 (2013)
20. Qiuxiang, F., Rong Y.: On the Lasota-Wazewska model with piecewise constant arguments. Acta Math. Scientia **26**, 371–378 (2006)
21. Richards, F.J.: A flexible growth function for imperical use. J. Exp. Bot. **10**, 290–301 (1959)
22. Rudin, W.: Principle of Mathematical Analysis, 3rd edn. McGraw Hill, New York (1976)
23. Smith, H.: An Introduction to Delay Differential Equations with Applications to the Life Sciences. Springer, New York (2010)

Printed in the United States
By Bookmasters